With kind regards

Anjo Stik
25/5/89

CHEMICAL PORPHYRIA IN MAN

CHEMICAL PORPHYRIA IN MAN

The diagnosis and occurrence of chronic hepatic porphyria in man caused by halogenated aromatics (polybrominated biphenyls, polychlorinated biphenyls and 2, 3, 7, 8-tetrachlorodibenzo-p-dioxin)

Porphyrinogenic action of halogenated aromatics in experimental animals

Edited by
J.J.T.W.A. STRIK
and
J.H. KOEMAN
Department of Toxicology
Agricultural University of Wageningen

1979

ELSEVIER/NORTH-HOLLAND BIOMEDICAL PRESS
AMSTERDAM · NEW YORK · OXFORD

ISBN: 0-444-80159-6

Published by:
Elsevier/North-Holland Biomedical Press
335 Jan van Galenstraat, P.O. Box 211
Amsterdam, The Netherlands

Sole distributors for the USA and Canada:
Elsevier North Holland Inc.
52 Vanderbilt Avenue
New York, N.Y. 10017

Library of Congress Cataloging in Publication Data

Main entry under title:

Chemical porphyria in man.

Bibliography: p.
1. Porphyria. 2. Halocarbons--Toxicology.
3. Aromatic compounds--Toxicology. 4. Liver--Diseases.
I. Strik, J. J. T. W. A. II. Koeman, J. H.
RC632.P6C47 616.3'99 79-19486
ISBN 0-444-80159-6

Printed in The Netherlands

PREFACE

Hepatic porphyria in man is a disorder of porphyrin metabolism which can either be inherited as a congenital anomaly or be caused by exposure to certain chemical compounds.

Among these so-called porphyrinogenic chemicals, which all belong to the group, the halogenated hydrocarbons, one finds a number of substances which are well known for their common use in industry and agriculture or because of their reputation as environmental contaminants, e.g. vinylchloride, hexachlorobenzene, certain brominated and chlorinated biphenyls and tetrachlorodibenzodioxin.

In clinical cases in man the condition of hepatic porphyria is usually associated with a considerable increase in the total amounts of porphyrins, especially uroporphyrin, excreted in the urine. However, in recent years it has been discovered that qualitative changes in the pattern of porphyrin metabolites in the urine provides a far more sensitive indication of these disturbances in the pathway of haem synthesis than the quantitative changes. One feature then is that, at total levels of excretion which fall within the normal range, the proportion of porphyrin metabolites containing eight and seven carboxylic groups is increased relative to those containing six, five and four of these groups. There is evidence that one of the key processes involved at the molecular level is the inhibition of the enzyme Uroporphyrinogen decarboxylase. The ability to disturb the function of this membrane bound enzyme thus seems to be a predominant property of porphyrinogenic chemicals. In this connection it is also important to mention that the drug enzyme induction and induction of hepatic porphyria very probably preceed or coinceed all other known effects which may be induced by these chemicals after subacute or short term exposure, such as neurological and hepatic morphological disorders.

The shift in the urinary porphyrin pattern does not appear to be related to any symptoms of disease either in animals exposed under experimental conditions or in human subjects accidentally exposed to a porphyrinogenic chemical. It is also found that this effect is completely reversible since the pattern is restored the normal as soon as the compounds are eliminated from the body. We are inclined therefore to consider this symptom as a pretoxic effect.

The routine measurement of porphyrin patterns may therefore represent a valuable parameter for the purpose of human biological monitoring in the occupational environment as well as under circumstances of gross environmental contamination like the PBB case in Michigan and the TCDD cloud in Seveso. It may even replace some of the parameters currently included in monitoring programmes such as certain serum enzymes like transaminases which certainly can not be considered as 'early' indicators of liver damage.

It is most likely that more porphyrinogenic halogenated hydrocarbons will be discovered since many of these compounds have not yet been checked for this property.

ACKNOWLEDGEMENT

The typing of this volume by Mrs. A.W. Wigman-van Brakel of the Central Typing Department of the Agricultural University is greatly appreciated.

J.H. KOEMAN

Department of Toxicology, Agricultural University, Wageningen, The Netherlands.

CONTENTS

LIST OF ABBREVIATIONS

Δ-A	Δ-absorbance
A_s	Absorbance (soret)
AA	amino acid transferase
AC	allylchloride
AHH	aryl hydrocarbon hydroxylase
ALA	δ-aminolevulinic acid
ALAS	δ-aminolevulinic acid synthase
BMC	bone marrow cell
BSP	bromosulphtaleine
C = CP = COPRO	coproporphyrin
CDD	chlorodibenzodioxin
CHP	chronic hepatic porphyria
Creat.	creatinine
DAB	Deutsches Arzneimittelbuch
DDC	3,5-diethoxycarbonyl-1,4-dihydrocollidine
DDE	1,1-di(p-chlorophenyl)-2,2-dichloroethylene
DDT	1,1-di(p-chlorophenyl)-2,2,2-trichloroethane
D-GA	D-glucaric acid
DM	diethylmaleate
ECH	epichlorohydrin
EPP	erythropoietic protoporphyria
F	female
GF	griseofulvin
GOT	glutamic oxaloacetic transaminase
GPT	glutamic pyruvic transaminase
γ-GT	γ-glutamyl transpeptidase
GSH	glutathione
HBB	hexabromobiphenyl
HCB	hexachlorobenzene
HCCP	hexachlorocylcopentadiene

H.D.	hydrolytic dechlorination
Ht	hematocrit
K_s	spectral dissociation constant
LDH	lactic dehydrogenase
M	male
MFO	mixed function oxidase
PAPS	3'-phosphoadenylsulphate
P = PP = PROTO	protoporphyrin
PB	phenobarbital
PBB	polybrominated biphenyl
PBG	porphobilinogen
PCB	polychlorinated biphenyl
PCP	pentachlorophenol
PCT	porphyria cutanea tarda
PCTA	pentachlorothioanisol
PCT-O	1-methyl(2,3,4,5,6-pentachlorophenyl)sulfoxide
$PCT-O_2$	1-methyl(2,3,4,5,6-pentachlorophenyl)sulfone
PCThP	pentachlorothiophenol
PeCB	pentachlorobenzene
PHA	polyhalogenated aromatic
ppb	parts per billion
PVC	polyvinylchloride
RBC	red blood cell
RBBC	red blood cell count
R.D.	reductive dechlorination
SER	smooth endoplasmic reticulum
SKF 525A	β-diethylaminoethyl-diphenylpropylacetate
TCB	tetrachlorobenzene
TCDD	2,3,7,3-tetrachlorodibenzo-dioxin
TCH	tetrachlorohydroquinone
TCP	trichlorophenol

TCTA	tetrachlorothioanisol
TCdi-TA	tetrachlorodithioanisol
TCThP	tetrachlorothiophenol
TLC	thin layer chromatography
TP	total porphyrins
U = UP = URO	uroporphyrin
UDPG	uridine diphospho glucuronate
UDPGA	uridine diphospho glucoronic acid
URO-D	uroporphyrinogen decarboxylase
UV	ultra violet
VC	vinylchloride
WBCC	White Blood Cell Count

INTRODUCTORY

Chemical Porphyria in Man, J.J.T.W.A. Strik and J.H. Koeman eds.

THE OCCURRENCE OF CHRONIC HEPATIC PORPHYRIA IN MAN CAUSED BY HALOGENATED HYDROCARBONS

J.J.T.W.A. STRIK
Department of Toxicology, Agricultural University
Biotechnion, De Dreijen 12, Wageningen (The Netherlands)

SUMMARY

Some of the polyhalogenated aromatic compounds (PHA's) which are able to produce chronic hepatic porphyria in experimental animals are presently known as environmental contaminants and have been involved in human poisoning episodes. One of the first measurable signals for human exposure to PHA's is alteration of the urinary porphyrin pattern.

INTRODUCTION

Definition and development of chronic hepatic porphyria

Chronic hepatic porphyria is a malfunction of the liver in which excess porphyrins are produced and excreted. The causes may be endogenous (Doss, this issue) or exogenous. Some polyhalogenated aromatic compounds are able to produce chronic hepatic porphyria both in man and in experimental animals (tabel 1). The presence of porphyrinogenic agents in human and animal food, in the work place, and in the environment poses a potential health hazard[1,2].

In the most severe form of porphyria, Porphyria Cutanea Tarda (PCT), a diagnostic indicator is the simultaneous significant increase of both uro- and heptacarboxylicporphyrin in urine. PCT is present when 45 to 80 % of the total porphyrins (1,500 to 13,000 µg/1 total porphyrins) is uroporphyrin and 15 to 35 % is heptacarboxylicporphyrin. This relative distribution pattern of the urine porphyrins is so characteristic that diagnosis can be made from a single examination of a small urine sample, even without knowledge of the 24-hour urine volume[3] (table 2).

In studies of the porphyrins in liver biopsies and urine, it has been found that chronic hepatic porphyria without clinical symptoms begins with accumulation of uro- and heptacarboxylicporphyrin in the liver, followed by a gradually increasing, pathologically elevated excretion of the porphyrins in the urine (200 - 1400 µg/l).

According to the type of porphyrin accumulation in the liver, the extent of porphyrinuria, and the relative distribution of uroporphyrin, coproporphyrin

TABLE 1

PORPHYRINOGENIC POLYHALOGENATED HYDROCARBONS

Halogenated benzenes	
Hexachlorobenzene	
Octachlorostyrene	
1,4-Dichlorobenzene	
1,2,4-Trichlorobenzene	
Hexabromobenzene	
Halogenated biphenyls	
2,4,5,2',4',5'-Hexachlorobiphenyl	Phenoclor DP6
2,3,4,2',3',4'-Hexachlorobiphenyl	Chlophen A60
3,4,5,3',4',5'-Hexachlorobiphenyl	Aroclor 1260
2,3,4,2',4',5'-Hexachlorobiphenyl	Aroclor 1254
Firemaster BP-6	Aroclor 1242
	Aroclor 1016
2,3,7,8-Tetrachlorodibenzo-p-dioxin	
γ-Hexachlorocyclohexane	
Halogenated aliphatics	
Mehtylchloride (?)	Vinylchloride (?)
	Allylchloride (?)

(?) not proven in hepatic tissue

and porphyrins with five to seven carboxylic groups in the urine, the chronic hepatic porphyrias (CHP) are divided into types A, B and C. These CHP-types are considered to be subclinical PCT. Type A, the mildest form, marks the beginning of the progression. The ratio of copro- to uroporphyrin is greater than 1. 5-15 % of the total urinary porphyrins is heptacarboxylic porphyrin.

Physiologically, the ratio of copro- to uroporphyrin is about 2:1 to 6:1. Type B is characterized by inversion of the normal ratio of coproporphyrin to uroporphyrin in the urine, with URO as the dominant porphyrin. 15 - 20 % of the total porphyrins is heptacarboxylic porphyrin. Chronic Hepatic Porphyria type C is latent PCT. 20 - 30 % of the total porphyrins is heptacarboxylic porphyrin. The biochemical signs of PCT are fully developed. PCT is defined by Doss[3,4]

TABLE 2

DISTRIBUTION OF PORPHYRINS AND TOTAL PROPHYRINS IN URINE IN HEPATIC PORPHYRIAS[4,5,21]

Chronic Hepatic Porphyria (CHP)	c/u-ratio	u (%)	7 (%)	Components	Total Porphyrins in urine (μg/l)
Normal	2-6	15-50	< 3	c>u>>>(traces of 7,6, 5)	0-200*
Coproporphyrinuria	> 6	15-50	3	c>u>5>7>6	100-200
Type A_1	> 6	15-50	5-15	c>u>5>7>6	200-600
Type A_2	> 6	15-50	5-15	c>u>7>5>6	200-600
Type B	< 1	> 50	15-20	u>c>7>5>6	200-600
Type C	<< 1	> 50	20-30	u>7>c>5>6	400-1400
Type D (Porphyria Cutanea Tarda) (PCT)	<<< 1	45-80	25-35	u>> 7>>>c (5,6)>5(c,6) >6(c,5)	600-1500

*This volume

Abbreviations: u and c = uro- and coproporphyrin,
7,6 and 5 = hepta-, hexa-, and pentacarboxylic porphyrins.

as clinical overt CHP. 25 - 37 % is heptacarboxylic porphyrin.

A chronic hepatic porphyrin synthesis disturbance is believed to develop in the following sequence: liver damage secondary porphyrinuria (coproporphyrinuria coprouroporphyrinuria) chronic hepatic porphyria (Type A_1 Type A_2 Type B) chronic hepatic porphyria Type C (latent Porphyria Cutanea Tarda) Porphyria Cutanea Tarda[5].

Chemically-induced porphyria in man

Historically, chemically-induced or acquired porphyria has been demonstrated in humans exposed via food supply or exposed in the workplace.

Hexachlorobenzene. Hexachlorobenzene (HCB) became widely known as the cause of a mass poisoning in Turkey due to the consumption of wheat treated with the fungicide (0.1 %). During the years 1955 to 1959 over 3,000 children were affected. Besides blisters and sores on the hands and face, dark pigmentation, and hairyness, these children usually had enlarged livers, and liver function tests showed marked impairment of liver function. The children were generally undersized and were apparently undernourished before they developed porphyria. Their dark-red or brown urine contained large quantities of porphyrins. There was a 10 % mortality rate[6]. Even after 10 years, many of those who survived still had disturbed liver functions and they showed cutaneous symptoms of porphyria in summer (table 3)[7]. In 1973 a case of porphyria in one worker out of 54 occupationally exposed to HCB used on grain was reported[8].

Polychlorinated biphenyls. In 1968 15,000 persons in Japan were affected more or less with jaundice, chloracne and dark coloring of the skin[9]. Porphyria was not investigated. These symptoms were attributed to a polychlorinated biphenyl (PCB's) contaminant in rice oil. Ten years later no porphyria could be detected (Strik et al., this volume and Nonaka et al., this volume).

Polybrominated biphenyls. In the summer and autumn of 1973, an unknown quantity of an industrial chemical mixture, consisting mostly of polybrominated biphenyls (PBB), entered Michigan's food supply. The mixture was sent to various feed mills where it was confused with magnesium oxide and added to dairy ration. Tens of thousands of cattle and other domestic animals consumed feed contaminated in one or more ways[10]. Millions of Michigan residents ate the meat and dairy products from these exposed animals, but it is presumed that the farmers who ate their own products probably received the highest doses of PBB.

PBB has been shown to be strongly porphyrinogenic in laboratory animals[11]. The studies reported (Strik et al., this volume) were undertaken to determine if porphyria could be found in a population suspected to have received the highest doses of PBB. Chronic hepatic porphyria type A could be found in members of farm families from Michigan.

Chlorodibenzodioxins. Chloracne and porphyria were reported in workers in 2,4,5-T plants (table 3)[12,13].

The impurity tetrachlorodibenzodioxin (TCDD) was the causative agent[14].

Chloracne due to industrial exporsure to TCDD has been known for a long time. Cases were reported in 1892, 1949, 1953, 1954/55, 1963, 1964, 1965-1966, 1968, 1970 and 1976.

In July 1976 in Seveso (Italy), environmental pollution by TCDD released

TABLE 3

DIAGNOSIS OF PORPHYRINS IN HUMAN URINE AFTER EXPOSURE TO PORPHYRINOGENIC POLYHALOGENATED AROMATIC AND POLYALIPHATIC COMPOUNDS

Compound	Number of persons	Total porphyrins (μg/24 hr) in urine	Porphyrin pattern	Fluorescence of porphyrins	Diagnosis
chloro-dibenzo dioxins	11	elevated[a]		urine	Porphyria Cutanea Tarda
	11	172-2230[b]		liver	
	3		URO,7		
hexa-chloro-benzene	5	3648-8778		urine ether-acid extract	Porphyria Cutanea Tarda
methyl-chloride	1	1000	only COPRO trace of URO		Porphyrinuria
vinyl-chloride	30	81-316[c]	URO 7,6,5, 3[d]		Secondary coproporphyrinuria and Chronic Hepatic Porphyria type A
	2		URO 7,5		
	1		URO,7		

[a]Watson-Schwartz test [b]uro- and coproporphyrin in normal range

[c]7,6,5,3 = 7-, 6-, 5-, 3-COOH porphyrin

references see table 1 (Strik, Doss, Schraa et al., this volume)

from a chemical factory was the cause of death of animals and interference with the health of the inhabitants. Chronic hepatic porphyria type A in people living in the contaminated area was detected in 1978 (Centen et al., this volume).

Vinyl- and methylchloride. Vinylchloride may cause an industrial disease involving chronic liver damage, scleroderma like cutaneous changes, thrombocytopenia, coproporphyrinuria and chronic hepatic porphyria (table 3)[15]. The occurrence of large amounta of coproporphyrins in urine in a case of industrial methylchloride poisoning is reported (table 3)[16]. Allylchloride seemed to be porphyrinogenic in experimental animals. Exposure of factory workers to allylchloride did not alter urinary porphyrin excretion (Nagelsmit et al., this volume).

DISCUSSION

Polyhalogenated hydrocarbons (HCB, 1,4 dichlorobenzene, 1,2,4-trichlorobenzene, octachlorostyrene, PCB, hexachlorobiphenyl isomers, TCDD, PBB, γ-hexachlorocyclohexane, methylchloride, vinylchloride and allylchloride have been shown to be porphyrinogenic in experimental animals, and, in some cases, in man[1 and this issue]. A feature of the porphyria evoked by these halogenated aromatics is the slow onset. A chronic exposure is usually needed[17]. Small birds, e.g., Japanese quail, are often used in laboratory experiments because of their sensitivity to the action of certain halogenated aromatic compounds and the rapidity with which they develop porphyria[18].

Hexachlorobenzene induced porphyria in the rat provides a suitable experimental model for studying the stages of development of human chronic hepatic porphyria[19]. Just as in chronic hepatic porphyria in man, the development of experimental hexachlorobenzene porphyria in the rat can be divided into several stages, chronic hepatic porphyria types A, B, C and D. In hexachlorobenzene porphyria in rats chronic hepatic porphyria type A develops after a relatively long interval of secondary coproporphyrinuria. The primary enzymatic defect in hexachlorobenzene porphyria in rats[19] as well as in the chronic hepatic porphyria in man is the diminished decarboxylation of uroporphyrinogen in the liver[20] indicating diminished uroporphyrinogen synthase activity. The increase of the excretion of uro- and heptacarboxylicporphyrin in the urine and the accumulation of both these porphyrins in the liver reflect directly the specific enzymatic defect in experimental and human chronic hepatic porphyria. The chronic hepatic porphyrias are the most common hepatic porphyrias in humans in Europe[21], and they are associated with various and different hepatic lesions, such as fibrosis, chronic aggressive hepatitis, fatty liver, siderosis and cirrhosis. For example, all stages of chronic hepatic porphyria, from type A up to type D, are found in liver cirrhosis of various pathogeneses[22].

As a group, polyhalogenated aromatics are very stable, persistent compounds. They resist degradation and accumulate in the food chain, concentrating ultimately in the fatty tissues of fish eating birds[23] and man. Their long term action results in liver cell damage. One of the first measurable signals for this liver damage is the alteration of porphyrin pattern and the increase of porphyrins in liver and urine.

REFERENCES

1. Strik, J.J.T.W.A. (1978) Porphyrinogenic Action of Polyhalogenated Aromatic Compouns with Special Reference to Porphyria and Environmental Impact, in M. Doss (Ed.), Proc. Int. Symp. Clin. Biochem. Diagnosis and Therapy of

Porphyrias and Lead Intoxication, Springer-Verlag, Berlin.

2. Strik, J.J.T.W.A. and Wit, J.G. (1972) TNO-Nieuws, 27, 604.

3. Doss, M. (1974) Porphyrins and Porphyrin Precursors, in M.C. Curtius and M. Roth (Eds.), Clinical Biochemistry, Principles and Methods, Vol. II, De Gruyter, Berlin, p. 1339.

4. Doss, M. (1975) Clinical Biochemistry and Regulation of Porphyrin Metabolism, in M. Doss (Ed.), Proc. German-Brazilian Seminar on Medicine and Biomedicine, Deutscher Akademischer Auschtauschdienst (DAAD), Bonn - Bad Godesberg, p. 13.

5. Doss, M. (1971) Klin. Wschr., 49, 939.

6. Cam, C. and Nigogosyan, G. (1963) J. Amer. Med. Assoc., 183. 88.

7. Dean, G. (1971) The Porphyrias, a Story of Inheritance and Environment, Pitman Medical, London.

8. Morley, A., Geary, D. and Harben, F. (1973) Med. J. Austr., 1, 565.

9. Kuratsune, M., Yoshimura, T., Matsuzaka, J. and Yamaguchi, A. (1972) Environm. Health Persp., april, 119.

10. Robertson, L.W. and Chynoweth, D.P. (1975) Environment, 17, 25.

11. Strik, J.J.T.W.A. (1973) Meded. Rijksfac. Landbouwwet. Gent, 38, 709.

12. Bleiberg, J., Wallen, M., Brodkin, R. and Applebaum, J.L. (1964) Arch. Derm., 89, 793.

13. Jirásek, L., Kalensky, J., Kubeck, K., Pazderová, J. and Lukás, E. (1976) Hautarzt, 27, 328.

14. Poland, A., Smith, D., Metter, C. and Possick, P. (1971) Arch. Environ. Health, 22, 316.

15. Lange, C.E., Block, H., Veltman, G. and Doss, M. (1976) Urinary Porphyrins among PVC Workers, in M. Doss (Ed.), Porphyrins in Human Diseases, Karger, Basel, p. 352.

16. Chalmers, J.N.M., Gillam, A.E. and Kench, J.E. (1940) Lancet, 28 dec., 806.

17. Strik, J.J.T.W.A. (1973) Enzyme, 16, 224.

18. Strik, J.J.T.W.A. (1973) Experimental Hepatic Porphyria in Birds, Thesis, Utrecht.

19. Doss, M., Schermuly, E. and Koss, G. (1976) Ann. Clin. Res., 8 (suppl. 17), 171.

20. Doss, M., Schermuly, E., Look, D. and Henning, H. (1976) Enzymatic Defects in Chronic Hepatic Porphyrias, in M. Doss (Ed.), Porphyrins in Human Diseases, Karger, Basel, p. 286.

21. Doss, M., Look, D., Henning, H., Lüders, C.J., Dölle, W. and Strohmeijer, G. (1971) Z. Klin. Chem. u. Klin. Biochem., 9, 471.

22. Doss, M., Look, D., Henning, H., Nawrocki, P., Schmidt, A., Dölle, W., Korb, G., Lüders, C.J. and Strohmeijer, G. (1972) Klin. Wschr., 50, 1025.

23. Koeman, J.H., van Velzen-Blad, H.C.W., de Vries, R. and Vos, J.G. (1973) J. Reprod. Fertil., Suppl. 19, 353.

Chemical Porphyria in Man, J.J.T.W.A. Strik and J.H. Koeman eds.

CHRONIC HEPATIC PORPHYRIAS IN HUMANS (ENDOGENIC FACTORS)

M. DOSS
Department of Clinical Biochemistry, Faculty of Medicine of the Philipp University, Marburg, Fed. Rep. Germany

SUMMARY

Chronic hepatic porphyrias, accompanied by chronic liver damage, exist in clinically latent and manifest stages, which can be recognized and differentiated on the basis of the characteristic urinary porphyrin excretion, porphyrin fluorescence of liver biopsy material and the accumulation of uro- and heptacarboxylicporphyrin in the liver. A stepwise progression appears to exist from the secondary coproporphyrinurias through clinically asymptomatic chronic hepatic porphyrias of Types A, B and C to clinically manifest Type D (porphyria cutanea tarda). A diminished activity of uroporphyrinogen decarboxylase in the liver is considered to be the primary enzymatic defect in chronic hepatic porphyrias. The most frequent observations were alcohol-toxic liver damage, chronic aggressive hepatitis and cirrhosis combined with chronic hepatic porphyrias. Alcohol and estrogens often lead to the clinical cutaneous manifestation of a chronic hepatic porphyria as 'porphyria cutanea tarda'.

The hepatic uroporphyrinogen decarboxylase defect seems to be a hereditary disturbance, since only about six percent of patients with liver disease develop chronic hepatic porphyria. In further studies uroporphyrinogen decarboxylase activity in red blood cells of both "sporadic" latent and manifest chronic hepatic porphyria patients and families with chronic hepatic porphyria was found to be depressed by about 40 %. These findings lend support to the hypothesis that chronic hepatic porphyria is a genetically determined disease, which only becomes manifest in combination with hepatopathic conditions induced by exogenous factors like alcohol and estrogens.

From family studies it can be concluded that the uroporphyrionogen decarboxylase defect is transmitted by an autosomal dominant trait.

The therapeutic and prophylactic management includes metabolic alkalisation, chloroquine, phlebotomy and diets rich in carbohydrates.

GENERAL REMARKS

The chronic hepatic porphyrias are metabolic disturbances caused by an abnormality in hepatic porphyrin synthesis. When in 1970 the term "chronic hepatic porphyria" (CHP)[1] was introduced, one also suggested at the same time a differentiation and differential diagnosis of the various stages of this chronic metabolic disease, the pathobiochemical origin of which is always connected with liver damage[2,3]. The chronic disturbance in porphyrin metabolism leads to skin symptoms in the final stage and was therefore called by clinicians "porphyria cutanea tarda"[4]. This definition of the disease "porphyria cutanea tarda", however, does not characterize the essence of the disease which is based on a disturbed porphyrin metabolism, it only describes it symptomatically: namely as a long term lasting skin disease due to a porphyrinopathy. The main feature of an endogen, genetically determined, or exogen, toxically determined disturbance of metabolism, is that this disturbance can only be revealed over a longer period of time by biochemical tests, therefore by changes in enzyme activity and metabolite concentrations. During these phases of clinical latency the clinician cannot make a diagnosis. A disturbance in metabolism, whether genetic or toxic, does not have to cause clinical signs, i.e. disturbances of clinical importance, a metabolic disease. The biochemical and clinical manifestation of a metabolic defect is on the one hand a question of variablity of gene penetration in endogen porphyrias and on the other hand a question of quality, quantity and duration of the toxical substances in exogen porphyrias. Exogenic factors which potentiate the biochemical disturbance of the defect, are of extreme importance for the clinical manifestation of genetically determined porphyrias: in acute hepatic porphyrias (table 1) they are a series of certain drugs, alcohol, estrogens and hunger; in chronic hepatic porphyrias they are liver damage, alcohol, estrogens and carcinoma.

The denotation "chronic hepatic porphyria" was based on three facts[1-3].

1. The centre of this disturbance is the liver[5,6].
2. The porphyria develops slowly, shows from the beginning a chronic course, which passes various characteristic subclinical phases, biochemically defined, and separable from each other[1-6].
3. Another reason for referring to the chronic hepatic disturbance in porphyrin metabolism which leads to "porphyria cutanea tarda" as "chronic hepatic porphyria" was then - and to a great extent still is the possibility of experimental imitation of this metabolic disturbance. Whereas so far a porphyria of the acute intermittent type could only be partially imitated, namely by inducing δ-aminolevulinic acid synthase with allylisopropylacetamide[7-9] it is possible, to a greater extent, to imitate a CHP in its biochemical characteristics by means of hexachlorobenzene[9,10], polyhalogenated biphenyls and also TCDD. These substances are also capable of producing toxic CHP in humans without a genetically predisposed uroporphyrinogen decarboxylase defect[11,12].

TABLE 1

CLASSIFICATION OF THE PORPHYRIN METABOLIC DISTURBANCES

I. Erythropoietic porphyrias
- A. Porphyria congenita erythropoietica (autosomal recessive)
- B. Protoporphyria erythrohepatica (autosomal dominant)

II. Hepatic porphyrias
- A. Porphyria acuta intermittens (autosomal dominant)
- B. Coproporphyria hereditaria (autosomal dominant)
- C. Porphyria variegata ("mixed hepatic porphyria", autosomal dominant)
- D. Chronic hepatic porphyrias ('constitutional' or hereditary: autosomal dominant; also paraneoplastic)
- E. Symptomatic toxic chronic hepatic porphyrias (acquired, as, for instance, "hexachlorobenzene-porphyria")

III. Acute lead poisoning (acute toxic porphyria; hepatic and erythropoietic)

IV. Secondary (symptomatic) coproporphyrinurias (acquired)
- A. Intoxications (alcohol, foreign chemicals, heavy metals, especially lead)
- B. Liver diseases (cirrhoses, hepatitis, siderosis; disorders of bile pigment metabolism)
- C. Blood diseases (hemolytic anemia, pernicious anemia, aplastic anemia, leukemia)
- D. Malignancies
- E. Cardiac infarction
- F. Infectious diseases
- G. Side effects of drugs

TABLE 2

CHRONIC HEPATIC PORPHYRIAS (CHP): BIOCHEMICAL AND CLINICAL MANIFESTATION

Condition	Constellation of urinary porphyrins	Total porphyrins (µmol/24 h)	URO %	HEPTA %	COPRO %	Hepatic porphyrin accumulation URO,HEPTA)	Liver damage	Cutaneous symptoms
Normal	C>>>U>>7≙5>6>3	<0.14	<20	<5	>70	-	-	-
Secondary Coproporphyrinuria	C>> U> 5≙7>3>6	<0.5	<20	<5	>70	-	(+)	-
CHP A	C > U> 7>5>6>3	<0.7	<40	<20	>40	(+)	+	-
CHP B	U > C> 7>5>6>3	<1.2	>40	<20	<40	+	+	-
CHP C	U > 7> C>5>6>3	<1.8	>50	<20	>10	++	+	-/(+)
CHP D	U > 7>>C>5>6>3	>1.9	>50	>20	<10	+++	+	+/++

Abbreviations: U and C uro- and coproporphyrin, 7, 6, 5 and 3 hepta-, hexa-, penta- and tricarboxylicporphyrin

DIFFERENTIATION OF CHRONIC HEPATIC PORPHYRIAS

The hepatic porphyrias represent by far the largest group of human porphyrin metabolic disturbances. Diagnosis is based mainly on assay of the porphyrins and their precursors in the urine, feces and liver[4-13].

Quantitation of uro- and coproporphyrins and the intermediate porphyrins with seven, six, and five carboxylic groups (hepta-, hexa-, and pentacarboxylicporphyrins) is of decisive importance for differentiation of the hepatic porphyrias, which develop typical patterns of porphyrin distribution in urine. The method of choice is thin-layer chromatographic separation followed by spectrophotometric or fluorimetric measurement[13]. In acute intermittent porphyria, both uro- and coproporphyrin can be the dominant components. In porphyria cutanea tarda, a diagnostic indicator is the simultaneous significant increase of uro- and heptacarboxylicporphyrin. Porphyria cutanea tarda is present when 45 - 80 % of the total porphyrins (1.5 to 15 mg/l total porphyrins) is uroporphyrin and 15 - 35 % is heptacarboxylicporphyrin. This relative distribution of urine porphyrins is specific for porhyria cutanea tarda[14]. The distribution patterns of the urine porphyrins is so characteristic of the various porphyrias that diagnosis can be made from a single examination of a small urine sample, even without knowledge of the 24-hour urine volume[2]. In studies of the porphyrins in liver biopsies and urine, it has been found that CHP's without clinical symptoms begin with accumulation of uro- and heptacarboxylicporphyrin in the liver, followed by a gradually increasing, pathologically elevated excretion of the porphyrins in the urine (table 2). Investigations on the porphyrin content of the liver and on renal porphyrin excretion in patients with histologically diagnosed liver diseases, e.g., fatty liver, fibrosis, chronic aggressive hepatitis, cirrhosis, and siderosis, have shown that it is possible to distinguish biochemically three types of asymptomatic CHP, which were designated as A, B, and C[1-3]. The first sign of CHP is the storage of uroporphyrin and later of heptacarboxylicporphyrin in the liver[1,5,6]. The diagram (fig. 1) is intended to illustrate how a chronic disturbance of hepatic porphyrin unfolds progressively by way of various biochemically distinct stages:

A liver lesion is capable of producing secondarily elevated excretion of coproporphyrin as well as CHP Type A in the presence of uroporphyrinogen decarboxylase defect. Increased excretion of uroporphyrin leads either to copro-uro-porphyrinuria (symptomatic) or to CHP type B (from Type A), which can also develop from secondary copro-uro-porphyrinuria after inversion of the copro-/uro-porphyrin ratio with concomitant increase of heptacarboxylicporphyrin. This inversion, together with a considerable elevation of excretion of uroporphyrin and heptacarboxylicporphyrin, must occur whenever coproporphyrinuria or

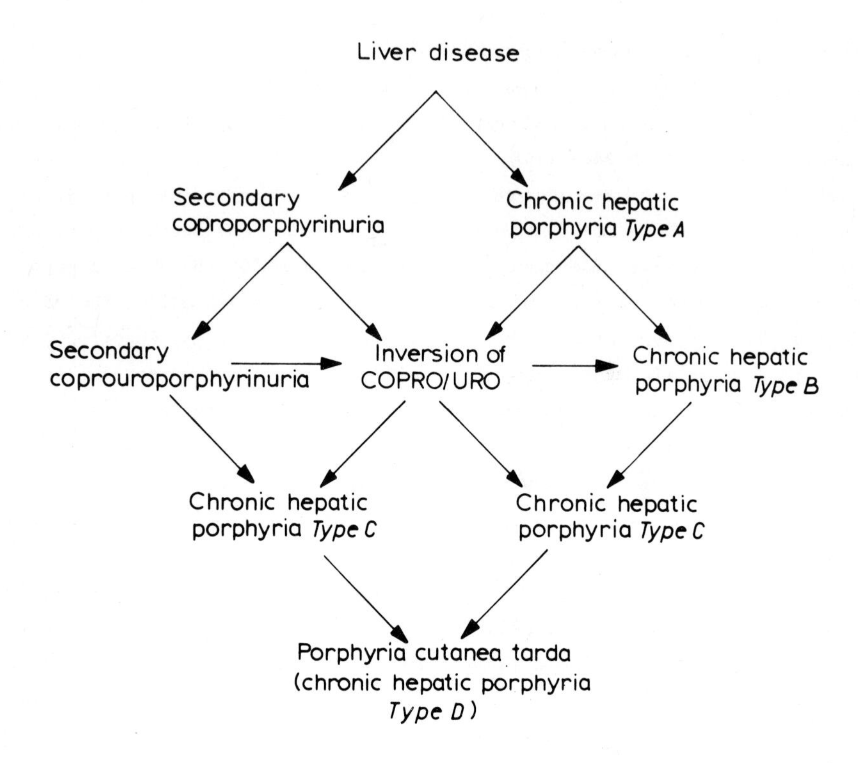

Fig. 1. Pathobiochemical development of acquired porphyrinurias and both inherited and acquired chronic hepatic porphyrias.
(Reproduced with permission of Springer Publishers, Berlin, New York)

copro-uro-porphyrinuria progresses into CHP Type C. The figure shows that porphyria cutanea tarda always develops from asymptomatic CHP Type C as the absolute amounts of uroporphyrin and heptacarboxylicporphyrin in the urine increase (table 2). Type C already exhibits the constellation of urinary porphyrins characteristic of porphyria cutanea tarda[2]. The first biochemical sign of disturbed hepatic porphyrin synthesis essential for the diagnosis of CHP is a moderate increase of uroporphyrin and then of heptacarboxylicporphyrin in the liver. The increase was observed only in the presence of liver damage, suggesting that the process leading to CHP unfolds primarily following injury to the liver, whereby endocrine and constitutional factors, as well as alcohol[15] and also certain drugs, sexual steroids[16], and foreign chemicals serve as aggravating factors[6,17]. As the condition develops from normal to porphyria cutanea tarda,

a continuous increase of uroporphyrin and heptacarboxylicporphyrin occurs in the liver, along with a gradual, continuous alteration of the coproporhyrin/uroporphyrin ratio in the urine[2,5,6]. The CHP's are divided into Types A, B, C and D. According to the type of porphyrin accumulation in the liver, the extent of porphyrinuria, and the relative distribution of uroporphyrin, coproporhyrin and porphyrins with five to seven carboxylic groups in the urine. In Type A, the ratio of copro- to uroporphyrin is greater than 1 (table 2). Physiologically, the ratio of copro- to uroporphyrin is about 2 : 6 : 1. In CHP Type B, the ratio of copro- to uroporphyrin is less than 1. CHP Type C is latent porphyria cutanea tarda. The biochemical signs of porphyria cutanea tarda are fully developed. A chronic hepatic porphyrin synthesis disturbance probably develops in the following sequence:
Liver damage (+ constitutional factors of hereditary factors) → secondary porphyrinuria (coproporphyrinuria copro-uro-porphyrinuria) → CHP (Type A → Type B) → CHP Type C → Porphyria cutanea tarda.

It was of interest to investigate whether the findings published in 1971[5] on a relatively small group of patients were in agreement with those obtained on a much larger collective, since reproducibility of the observations would lend considerable support to the contention that CHP is indeed a latent form of porphyria cutanea tarda.

The occurrence of CHP and the differentiation of this condition into Types A, B, C and D[3] were studied with regards to excretion of porphyrins in the urine of 289 patients hospitalized with various liver diseases[18]. Patients with even the slightest suspicion of disturbance in porphyrin metabolism were included (positive talc fluorescence test, liver fluorescence in laparoscopy, cultaneous symptoms even remotely suggestive of CHP). The CHP types represent stages in a continuous progression. All Type D patients had at least discrete cutaneous symptoms. Of the 289 patients, 106 had CHP; the rest had either symptomatic coproporphyrinuria or porphyrin excretion in the upper normal range. Differentiation of our CHP cases into types was based on the following constellations of uroporphyrin (U), heptacarboxylicporphyrin (7) and coproporphyrin (C) in the urine, all of which were elevated:

Symptomatic coproporphyrinuria	(n=123)	C>> U
CHP Type A	(n= 37)	C > U > 7
CHP Type B	(n= 20)	U > C > 7
CHP Type C	(n= 21)	U > 7>> C
CHP Type D	(n= 28)	U > 7>>>C

The excretion pattern of porphyrins in Types A to D compared with normals is depicted in fig. 2 based on absolute values (μg/24 hours), and in fig. 3, 4, 5, 6 and 7 based on relative distribution (%). For comparison, fig. 3 also contains previously published data on another group of patients[5].

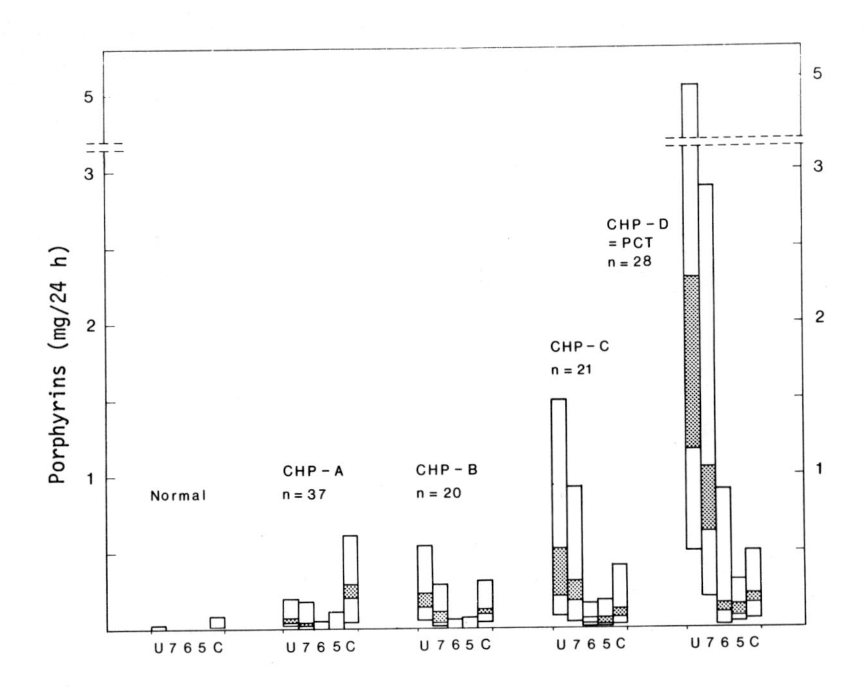

Fig. 2. Urinary porphyrin excretion pattern in chronic hepatic porphyria Types A - D (μg/24 hours)[18]. U = uroporphyrin, 7 = heptacarboxylic porphyrin, 6 = hexacarboxylic porphyrin, 5 = pentacarboxylic porphyrin, C = coproporphyrin. The whole columns represent the range (95 % of values). Due to the exponential scattering of the values, evaluation was based on logarithms. The shaded portions indicate the range in which the true median can be expected with a probability exceeding 95 %. For statistical evaluation of comparison of excretion of each porphyrin fraction in each group see ref. 18.

Significant differences among the Types A to D as determined by Student's t-test were found: The drop in proportion of coproporphyrin from Types A to C and D is highly significant, as is the relative and absolute increase of uroporphyrin[18]. The increase of heptacarboxylicporphyrin is continuous, but significant only on comparison of absolute excretion in Types A and D. Student's t-test revealed no significant differences between the data on the 32 patients

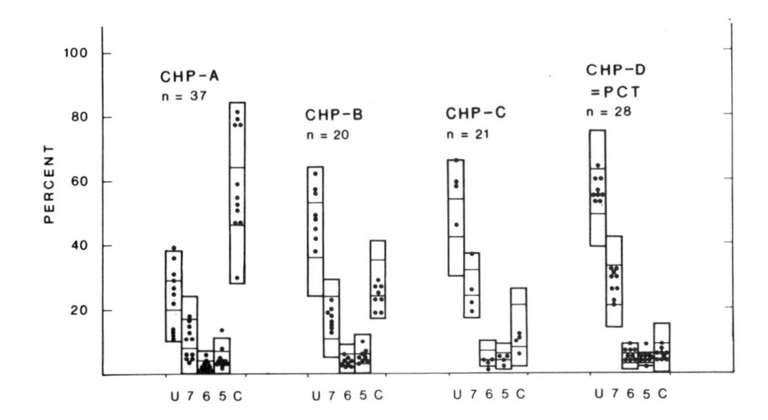

Fig. 3. Urinary porphyrin excretion pattern in chronic hepatic porphyria Types A - D[18]. Percent distribution in two groups, based on excretion in µg/24 hours. Key to U, 7, 6, 5, C see legend to Figure 2. Whole columns = range of own data (n = 106) (95 % of values, mean $\pm$ 2s); middle portions indicate the range in which the true median can be expected with a probably exceeding 99 %. Dots = individual values from ref. 5 (n = 32).

from 1971[5] and the present findings on absolute excretion and percent distribution of porphyrins in a much larger group.

These results show that a formal progression exists from CHP Type A through to Type D. They also demonstrate statistically significant differences in range and mean values among the four types of CHP, both regarding the absolute amounts and relative distribution of urinary porphyrins. Satisfactory reproducibility is indicated by comparison with a second, smaller collective. Thus the division into types of chronic hepatic porphyria proposed in previous papers[1-3,5,6] confirms to two basic requirements for clinical usefulness, statistical distinction of the categories and reproducibility.

PATHOBIOCHEMISTRY OF CHRONIC HEPATIC PORPHYRIAS

In chronic hepatic porphyria the pattern of excretion of porphyrins in the urine and their accumulation in the liver clearly suggest the presence of one or more enzymatic defects, which are most likely to be found in the decarboxylation of uro- and heptacarboxylic porphyrinogens. In chronic hepatic porphyrias including porphyria cutanea tarda the liver has a high initial concentration of porphyrin due to accumulation of uroporphyrin and heptacarboxylicporphyrin; 46 - 30 % of the uroporphyrin consists of isomer III, whereas heptacarboxylicporphyrin contains practically no isomer I. Heptacarboxylicporphyrin makes up

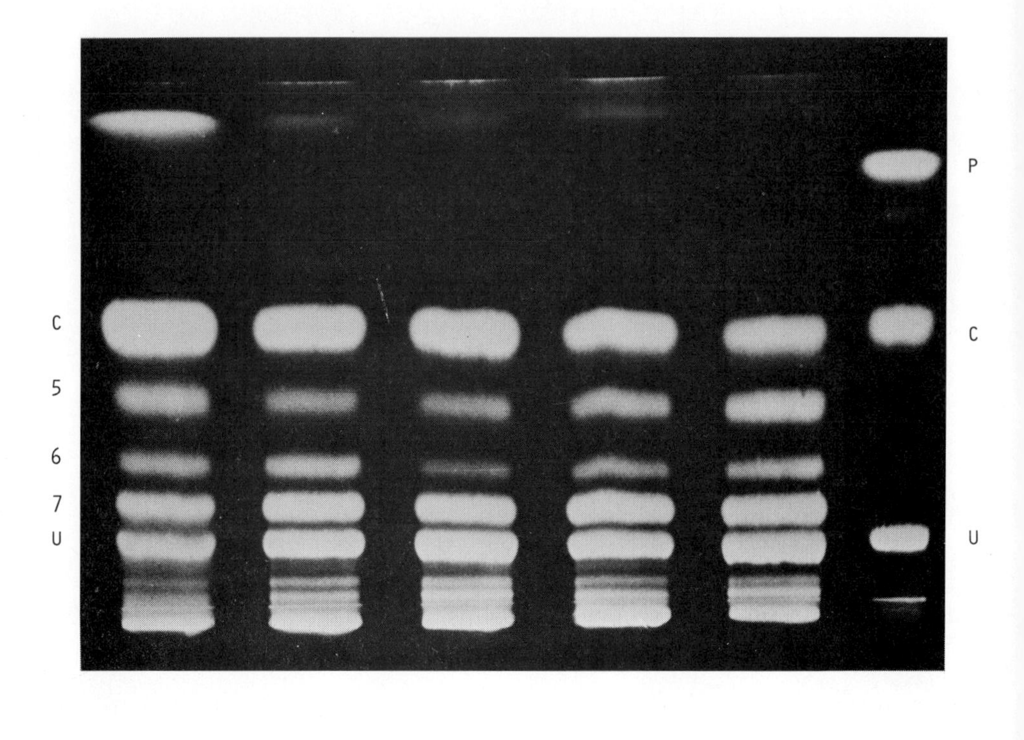

Fig. 4. Chromatographic profiles of urinary porphyrins from patients with chronic hepatic porphyria Type A, B, B, C, and D (from left to right), for experimental conditions see ref. 13). In all chronic hepatic porphyria stages an abnormal excretion of HEPTA is observed. In Type A COPRO, URO, and HEPTA increase, but URO becomes dominant. In the chronic hepatic porphyrias a progressive decrease of COPRO parallels a progressive increase in the level of HEPTA from Type A to Type C, in the course of which HEPTA moves from the fourth (Type A) to become the second most abundant porphyrin (Type C); its increase is extreme in Type D (porphyria cutanea tarda).

from 11 to 37 % of the porphyrins in the liver in chronic hepatic porphyria. The amount of heptacarboxylicporphyrin can be completely accounted for by assuming its derivation from uroporphyrinogen III. Liver homogenate from chronic hepatic porphyria patients produces mainly uroporphyrin and heptacarboxylicporphyrin from porphobilinogen[19]. In porphyria cutanea tarda liver homogenate the isomer composition of porphyrins after their synthesis from porphobilinogen is similar to that in nonincubated liver biopsy tissue: uroporphyrin III 36 % and heptacarboxylicporphyrin III 100 %. The low isomer III values seems to reflect

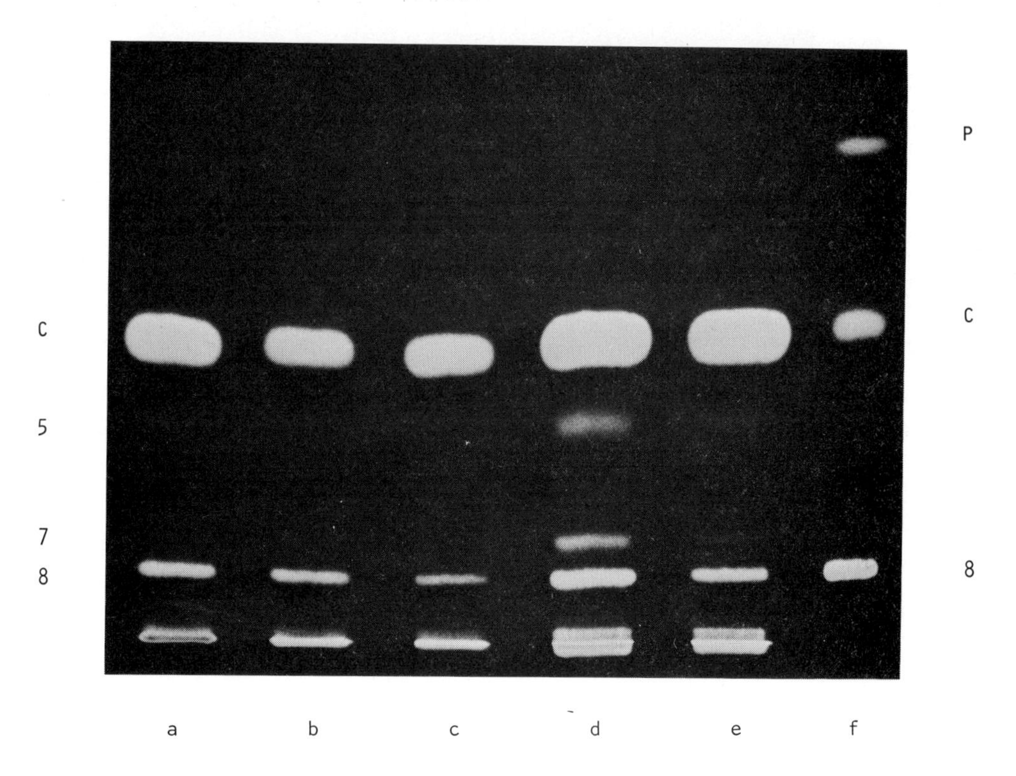

Fig. 5. Silica gel thin-layer chromatogram with the porphyrins from 10 ml urine samples from three normal subjects (a to c), and from two patients with alcohol fatty liver (d, e). The porphyrins were separated as the methyl esters in the solvent system benzene-ethyl acetate-methanol (85:13.5:1,5, by vol.). On the right side (f), comparison substances: uro-, copro-, and protoporphyrin methyl esters (U, C, and P). Urine porphyrins: uro-, heptacarboxylic and coproporphyrins (U, 7, and C). Photographed under the UV light from a mercury high pressure lamp.

a uroporphyrinogen cosynthase defect in this disease. The results from chronic hepatic porphyria livers indicate a decarboxylase defect with manifests itself in different decarboxylation steps in the isomer chain I and III (table 3). These defects may be responsible for the hepatic accumulation of uroporphyrin I (70 % of total uroporphyrin) and heptacarboxylicporphyrin III (100 %). The hepatic uroporphyrinogen decarboxylase defect is the primary enzymatic defect in chronic hepatic porphyrias[19].

The hepatic uroporphyrinogen decarboxylase defect seems to be a hereditary disturbance, since only about six percent of patients with liver disease develop chronic hepatic porphyria and, in addition, disturbed decarboxylation of uroporphyrinogen has been found in red cells of patients with hepatocellular

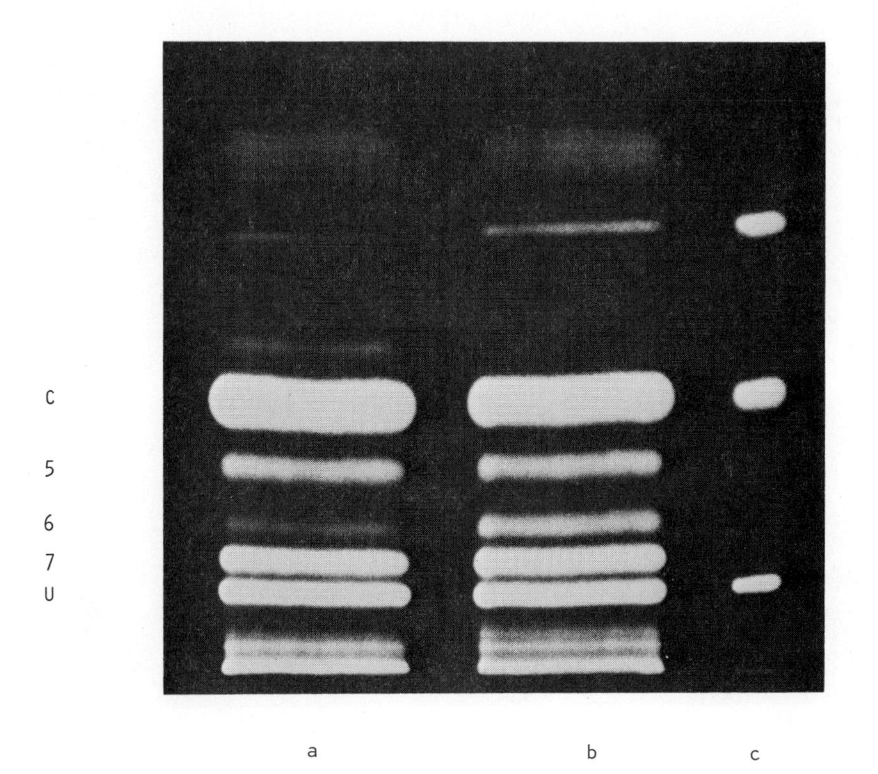

Fig. 6. Fluorescence record of urinary porphyrin TLC of two cases with chronic hepatic porphyria Type A (a and b) compared to reference substances (c). For abbreviations see legend to figure 6.

TABLE 3

HEPATIC ENZYMATIC DEFECTS IN CHRONIC HEPATIC PORPHYRIA

1 Uroporphyrinogen I	decarboxylase	heptacarboxylic porphyrinogen I
2 Heptacarboxylic porphyrinogen III	decarboxylase	hexacarboxylic porphyrinogen III
3 Uroporphyrinogen cosynthase		

Gradual development from CHP A up to CHP D (PCT)

and prostata carcinoma associated with porphyria cutanea tarda[20,21].

In further studies uroporphyrinogen decarboxylase activity in red blood cells of both "sporadic" latent and manifest CHP patients and families with chronic hepatic porphyria was found to be depressed by about 40 % (table 4)[22]. The activity of uroporphyrinogen synthase in red blood cells is elevated in chronic hepatic porphyria: This increase in uroporphyrinogen synthase activity may reflect a partial compensatory mechanism against the relative deficiency of uroporphyrinogen decarboxylase[23,24]. These findings on uroporphyrinogen decarboxylase in erythrocytes and urinary porphyrin excretion in several Italian CHP families agree with the observations here, lending support to the hypothesis that CHP is a genetically determined disease, but requires for manifestation in humans exogenous factors such as alcohol or estrogens in combination with hepatopathy (fig. 7). From family studies it can be concluded that the uroporphyrinogen decarboxylase defect is transmitted by an autosomal dominant trait[22,25]. The enzyme defect in the erythropoietic system is not known to produce any deleterious effects. Even in the liver the uroporphyrinogen decarboxylase defect does not seem to cause any lability of the control mechanisms in heme synthesis, in contrast to the acute forms of hepatic porphyria, which are essentially "regulatory disorders" at the molecular level[13].

On the one hand chronic hepatic porphyrias have a hereditary basis, but, on the other hand, they can also be induced in humans and animals without a pre-existing uroporphyrinogen decarboxylase defect by foreign toxic chemicals such as polyhalogenated hydrocarbons. In contrast to familiar "sporadic" and paraneoplastic chronic hepatic porphyria in humans with a genetically fixed underlying uroporphyrinogen decarboxylase defect, a lesion of this enzyme will

TABLE 4

UROPORPHYRINOGEN DECARBOXYLASE IN RED BLOOD CELLS OF CHP PATIENTS COMPARED TO CONTROLS[22]

		URO-D activity (COPRO µmol/l.h; $\bar{x} \pm s$)
CHP/PCT family members	(n = 27)	12 ± 4
CHP/PCT patients ("sporadic")	(n = 32)	11 ± 3
Total observations	(n = 59)	11 ± 3.5
(Depression of activity ≙ 45 %)		
Controls	(n = 120)	20 ± 3
Significance	(n = 120)	$p < 0.001$

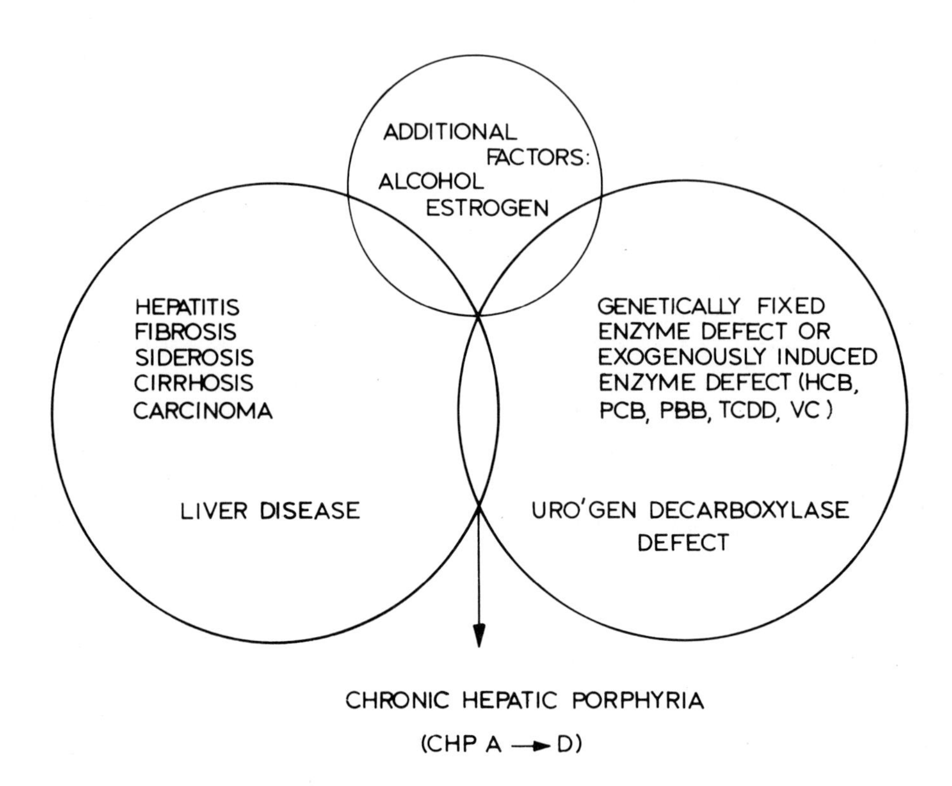

Fig. 7. Suggested pathobiochemical development of chronic hepatic porphyrias including the coincidence of endogenous and exogenous factors. Exclusively exogenous factors may be responsible in chronic hepatic porphyria cases induced by environmental chemicals.

be induced by hexachlorobenzene and similar substances, whereby the development of hexachlorobenzene porphyria in rats, as reflected by urinary and hepatic porphyrins, proceeds in stages analogous to those of CHP in man[9,10], and is also combined with a toxic liver injury. Therefore human and experimental CHP can also be regarded as "membrane disease"[26].

In the light of recent evidence on the enzyme defects in the porphyrias, it now seems doubtful that even the clinically sporadic cases of porphyria cutanea tarda are purely acquired. Like that of acute intermittent porphyria. the development of CHP can probably be divided into three stages: (a) the phase of the genetic enzymatic defect alone; (b) biochemical manifestation (pathologic excretion of metabolites of porphyrin synthesis and porphyrin accumulation in the liver); and (c) the clinical syndrome. Recognition of CHP in clinically symptom-free stages is of great usefulness to the patient, in that suitable measures can be initiated to stop or slow the progression to manifest porphyria cutanea tarda. In addition to treatment of the liver disease, these measures include absolute prohibition of alcohol, avoidance of porphyrinogenic drugs (especially estrogens), adherance to a high-carbohydrate diet, and possible metabolic alkalization[27].

ACKNOWLEDGEMENT

The experimental work of these studies was supported by the "Deutsche Forschungsgemeinschaft".

REFERENCES

1. Doss, M. (1970) Hoppe-Seyler's Z. physiol. Chem. 351, 1300.
2. Doss, M. (1971) Klin. Wschr. 49, 939.
3. Doss, M. (1971) Klin. Wschr. 49, 941.
4. Doss, M. and Meinhof, W. (1971) Dtsch. med. Wschr. 96, 1006.
5. Doss, M., Look, D., Henning, H., Lüders, C.J., Dölle, W. and Strohmeyer, G. (1971) Z. Klin. Chem. u. Klin. Biochem. 9, 471.
6. Doss, M., Look, D., Henning, H., Nawrocki, P., Schmidt, A., Dölle, W., Korb, G., Lüders, C.J. and Strohmeyer, G. (1972) Klin. Wschr. 50, 1025.
7. Doss, M. (ed.) (1974) Regulation of Porphyrin and Heme Biosynthesis, Karger, Basel.
8. Doss, M. (ed.) (1976) Porphyrins in Human Diseases, Karger, Basel.
9. Doss, M. and Nawrocki, P. (eds.) (1976) Porphyrins in Human Diseases - Report of the Discussions, Dr. Falk, Freiburg.
10. Doss, M., Schermuly, E. and Koss, G. (1976) Ann. clin. Res. 8 suppl. 17, 171.
11. Doss, M. and Schwartz, S. (1978) J. Clin. Chem. Clin. Biochem. 16, 25.
12. Doss, M. (ed.) (1978) Diagnosis and Therapy of Porphyrias and Lead Intoxication, Springer, Berlin-Heidelberg-New York.
13. Doss, M. (1974) Porphyrins and porphyrin precursors, in Clinical Biochemistry (H.-Ch.Curtius and M.Roth, eds.), Walter de Gruyter, Berlin - New York, Vol. II, 1323.
14. Doss, M., Meinhof, W., Malchow, H., Sodomann, C.-P. and Dölle, W. (1970) Klin. Wschr. 48, 1132,
15. Doss, M., Nawrocki, P., Schmidt, A., Strohmeyer, G., Egbring, R., Schimpff, G., Dölle, W. and Korb G. (1971) Dtsch. med. Wschr. 96, 1229.
16. Doss, M. (1977) Dtsch. med. Wschr. 102, 875.

17. Doss, M. (1977) Clinical biochemistry and regulation of porphyrin metabolism, in Proc. of the German-Brazilian Seminar on Medicine and Biomedicine (M. Doss, ed.), Deutscher Akademischer Austauschdienst (DAAD), Bonn-Bad Godesberg, 13.

18. Doss, M., Nawrocki, P. and Schermuly, E. (1976) Diagnostic profiles of urinary porphyrin precursors and porphyrin excretion in hepatic porphyrias and lead intoxication, in: Porphyrins in Human Diseases - Report of the Discussions (M. Doss and P. Nawrocki, eds.), Dr. Falk, Freiburg, 163.

19. Doss, M.. Schermuly, E., Look, D. and Henning, H. (1976) Enzymatic defects in chronic hepatic porphyrias, in: Porphyrins in Human Diseases (M. Doss, ed.), Karger, Basel, 286.

20. Doss, M., Tieperman, R.v., Schermuly, E., Scheffer, B. and Martini, G.A. (1976) Primary liver carcinoma associated with porphyria cutanea tarda and uroporphyrinogen decarboxylase defect in erythrocytes, in: Porphyrins in Human Diseases - Report of the Discussions (M. Doss and P. Nawrocki, eds.), Dr. Falk, Freiburg, 222.

21. Doss, M. and Martini, G.A. (1978) Porphyrin Metabolism and liver tumors, in: Primary Liver Tumors (H. Remmer and H.M. Bolt, eds.), MTP, Lancaster.

22. Tiepermann, R.v. and Doss, M. (1978) Uroporphyrinogen decarboxylase defect in chronic hepatic porphyria, in: Diagnosis and Therapy of Porphyrias and Lead Intoxication (M. Doss, ed.), Springer, Heidelberg-Berlin-New York.

23. Schermuly, E. and Doss, M. (1976) Ann. clin. Res. 8, suppl. 17, 92.

24. Doss, M. (1977) Med. Klin. 72, 1501.

25. Kushner, J.P., Barbuto, A.J. and Lee, G.R. (1976) J. Clin. Invest. 58, 1089.

26. Doss, M. (1977) Innere Med. 4, 5 and 43.

27. Doss, M. (1977) Internist 18, 664.

CASE STUDIES

Chemical Porphyria in Man, J.J.T.W.A. Strik and J.H. Koeman eds.

COPROPORPHYRINURIA AND CHRONIC HEPATIC PORPHYRIA TYPE A FOUND IN FARM FAMILIES FROM MICHIGAN (U.S.A.) EXPOSED TO POLYBROMINATED BIPHENYLS (PBB)

J.J.T.W.A. STRIK[a], M. DOSS[b], G. SCHRAA[a], L.W. ROBERTSON[c], R. von TIEPERMANN[b] and E.G.M. HARMSEN[a].

[a]Department of Toxicology, Agricultural University, Wageningen, The Netherlands.

[b]Department of Clinical Biochemistry, Faculty of Medicine of the Philipp University, Marburg a.d. Lahn, D-3550, F.R.G.

[c]Department of Environmental and Industrial Health, School of Public Health, The University of Michigan, Ann Arbor, Michigan 48109, U.S.A.

SUMMARY

In 1977 urine samples were collected from members of farm families in Michigan, U.S.A., who had ingested polybrominated biphenyl (PBB) - contaminated meat and dairy products beginning in 1973, and from members of farm families in Wisconsin with no known exposure to PBB (control group).

Total porphyrin excretion of both groups was below 200 µg/l. Therefore at this level this parameter cannot be used as an indication for exposure of humans to PBB.

Out of a group of 142 persons, at least 47 % with secondary coproporphyrinuria or with chronic hepatic porphyria Type A were found demonstrating an abnormal porphyrin pattern. The incidence of this indicator of liver malfunction was higher in this PBB-exposed group than in controls (6 %). This study is limited solely to an assessment of liver damage as manifested by porphyrin excretion.

INTRODUCTION

Chemically-induced porphyria in man

Historically, chemically-induced or acquired porphyria has been demonstrated in humans exposed via food supply or exposed in the workplace. In 1955 and 1956, consumption of wheat that had been treated with hexachlorobenzene caused porphyria in humans in Turkey[1]. In addition, porphyria has been demonstrated in workers exposed to tetrachlorodibenzo-dioxin (TCDD) - contaminated 2, 4, 5 - trichlorophenoxyacetic acid[2,3,4], methylchloride[5], and vinylchloride[6].

TABLE 1

DIAGNOSIS OF PORPHYRINS IN HUMAN URINE AFTER EXPOSURE TO PORPHYRINOGENIC POLYHALOGENATED AROMATIC AND ALIPHATIC COMPOUNDS

Compound	Number of Persons	Total porphyrins (μg/24 hr) in urine	Porphyrin pattern	Fluorescence of Porphyrins	Diagnosis	Reference
chlorodibenzo dioxins	11	elevated[a]		urine	Porphyria Cutanea Tarda	2
	11	172-2230[b]		liver		
	3		URO,7[d]			3
hexachlorobenzene	5	3648-8778		urine ether-acid extract	Porphyria Cutanea Tarda	1
methylchloride	1	1000	only COPRO, trace of URO		Porphyrinuria	5
vinylchloride	30 2 1	81-316[c]	URO,7,6,5,3 URO,7,5 URO,7		Secondary coproporphyrinuria	6

[a]Watson-Schwartz test

[b]Uroporphyrin, coproporphyrin in normal range

[c]Coproporphyrin

[d]7,6,5,3 = 7-, 6-, 5-, 3-COOH porphyrin

PBB accident

In the summer and autumn of 1973, an unknown quantity of an industrial chemical mixture, consisting mostly of polybrominated biphenyls (PBB), entered Michigan's food supply. The mixture was sent to various feed mills where it was confused with magnesium oxide and added to dairy ration. Tens of thousands of cattle and other domestic animals consumed feed contaminated in one or more ways[7]. Millions of Michigan residents consumed the meat and dairy products from these exposed animals, but it is presumed that the farmers who ate their own products probably received the highest doses of PBB[8].

PBB has been shown to be strongly porphyrinogenic in laboratory animals[9]. The study reported here was undertaken to determine whether porphyria occurs in a population suspected of having received the highest doses of PBB.

MATERIALS AND METHODS

Group selection

Sixty-one Michigan farm families were contacted by letter in May, 1977. These families were chosen because:

1) they lived on or received food from PBB-contaminated farms;
2) some members of each family were found to contain PBB in their blood or adipose tissue; and
3) they complained of health problems.

A follow-up letter was sent in June, 1977 explaining the sampling programme in more detail (e.g., that the first morning urine should be collected, that the collected urine should be kept cold and in the dark), and in late June, nalgene bottles and labels were mailed to each family. The bottles were filled and labelled with the name, age and date the sample was taken.

The filled bottles were collected either directly from the farmers or through the mail. For the most part the received bottles were frozen; those which were not, were cold. Immediatley upon collection the bottles were placed on dry ice, and transported to a food locker (-18 ^{o}C) where they were kept until they were shipped (in dry ice) to Wageningen. The samples arrived hard frozen and were held in a freezer (-50 ^{o}C) until used.

Group description

The farm families in most cases received fewer bottles than they had family members; each family chose which members would be sampled. The result was the selection of a group which was 60 % male and 60 % adult.

TABLE 2
AGE-SEX DISTRIBUTION OF THE PBB STUDY GROUP

	Male		Female		Group Totals	
Under 19	34(41)*	[62]	21(36)	[38]	59(39)	[100]
19 and older	49(59)	[56]	38(64)	[44]	87(61)	[100]
Total	83(100)	[58]	59(100)	[42]	142(100)	[100]

*For percentages read parentheses down and brackets across.

The number of blood and adipose tissue analyses which were available are tabulated in tabel 3, showing the mean blood level 3 ppb PBB and a mean adipose tissue level of 686 ppb PBB for the group.

TABLE 3

PBB BLOOD AND ADIPOSE TISSUE LEVELS (ppb)

	blood			adipose		
	$\bar{x}$	s	n	$\bar{x}$	s	n
Male	3	2	13	756	413	16
Female	3	2	9	687	1193	8
Total	3	2	22	686	720	26

Group selection of control group

With a great amount of help from Barbara L. Lear, P.A. a control group of 141 people was collected, consisting of farm families from Wisconsin. They were chosen because of their American dietary habits which closely resembled those of the people from Michigan and the fact that they were not exposed to PBB. Their first morning urine collected in the first two weeks of November 1977 was treated the same way as the urine from the Michigan farm families.

Group selection of resamplings

From the group sampled in June 1977 two individuals were contacted in November to provide a urine sample over 7 consecutive days. They were chosen because one had a high total porphyrin level, and the other one had a low level in their morning urine during the June sampling. Also 10 farm families were chosen at random from those who provided urine samples in June 1977. They were asked to provide one further morning urine sample. Of these families 8 responded (27 persons). Bottles containing urine were stored frozen.

Group selection of relatives of farm families from Michigan, who live outside the state

All farm families sampled in June were contacted by telephone in October. They were asked if they had relatives who lived outside the state of Michigan since 1973, the year in which the PBB accident took place. These persons were contacted by letter and received one to four bottles per family to provide morning urine samples. 32 Persons responded by returning the bottle. This group was used as a control group to see if any genetic factors might be involved in porphyrin excretion.

Group selection of PBB exposed individuals from which serum enzyme activities and urinary porphyrin data were known

46 Members of farm families from Michigan exposed to PBB and from whom the urinary porphyrin pattern had been determined were also included in the clinical field study of November, 4-10, 1976 by the Environmental Sciences Laboratory of the Mount Sinai School of Medicine, New York.

Besides a wide variety of parameters, which gave information on possible adverse health effects amongst individuals exposed to PBB, liver function tests were also used e.g. SGOT, SGPT and LDH.

Randomizing the samples

The samples were numbered and placed in a cold room (4 °C). A sample was then chosen for analysis, by a sample number being obtained from a random number table.

URINE PORPHYRIN ANALYSES

Measurement of Total Porphyrins in Urine was done According to Doss[10], this volume.

Recovery of total porphyrins

To test the recovery of porphyrins from the ion-exchange columns, 10 urine samples were analyzed for total porphyrins and these same samples were analyzed again after having been spiked with coproporphyrin-1 (stock No. COP-1-5, lot 678-C6, Sigma Chemical Company, P.O.Box 14508, St.Louis, Missouri).

The corresponding urines were run consequently on the same columns (unspiked were done first). The absorbances were corrected and the total porphyrin in the urine was calculated as before. The total porphyrin in the standard was calculated according to the following formula:

$$TP = \frac{A_{s(corrected)} \times 654 \times HCl\ volume\ (ml)}{428 \times d\ (cm) \times volume\ of\ the\ standard\ (ml)}$$

The value arrived at for the standard is the mean of 3 determinations. Recovery was calculated as follows:

$$Recovery = \frac{TP_{spiked\ urine} - TP_{urine}}{TP_{standard}}$$

The results of the recovery experiment (which were the 17th and 18th runs of ion exchange columns) are seen in table 4.

TABLE 4

RECOVERY OF PORPHYRIN ADDED TO URINE

	Total Porphyrin (µg/l)			
Sample	Urine	Porphyrin Added	Spiked Urine	Recovery
1	131	257*	325	.75
2	48	257	249	.78
3	42	257	163	.47
4	66	257	228	.63
5	83	257	225	.55
6	62	257	225	.63
7	31	257	187	.62
8	62	257	235	.67
9	80	257	221	.55
10	66	257	228	.63
				$\bar{x}$ = .63
				s = .09

*Average of 3 determinations

The mean recovery efficiency was calculated to be 63 %.

Thin Layer Chromatography of the porphyrins from urine as methyl esters was carried out according to Doss[10], this volume

Creatinine was measured according to Gorther and de Graaff[11]. This was done to eliminate the factor of different urine concentrations.

RESULTS

Total porphyrin values

In figures 1 and 2 the Michigan and Wisconsin groups are subdivided into males, females, adults and young persons.

No significant differences between urinary total porphyrin excretion of different groups are found.

In November 1977, morning urine was re-collected, over 7 consecutive days, from one member of the Michigan group. These results and the values found in his morning urine from June 1977 are shown in table 5. The data fluctuate from 76 up to 165 µg/l or 7.1 up to 19.7 µg/mmol creatinine.

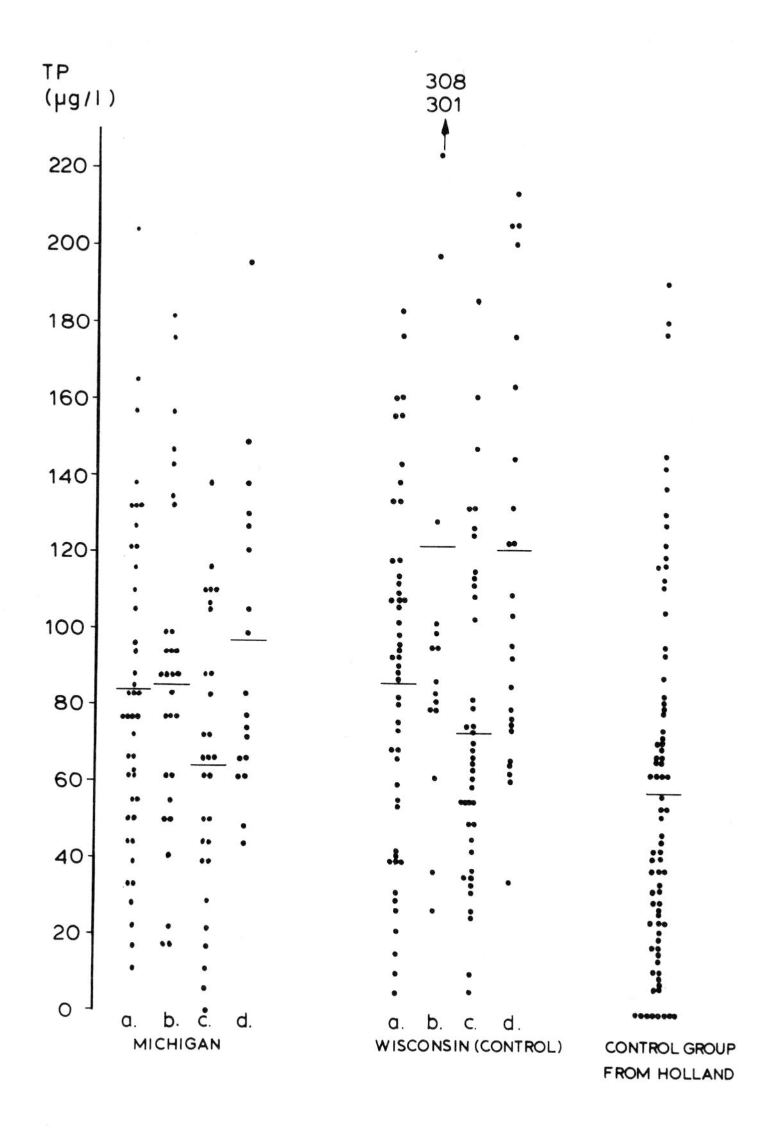

Fig. 1. Total porphyrin values (μg/l) in the Michigan and Wisconsin groups, subdivided in males, females, adults and young people and in a Dutch group

	Michigan			Wisconsin		
	$\bar{x}$	s	n	$\bar{x}$	s	n
male , ≥ 19 years(a)	84	43	45	87	45	53
male , < 19 years(b)	85	45	31	122	82	17
female, ≥ 19 years(c)	64	34	31	73	43	42
female, < 19 years(d)	97	40	19	121	60	25

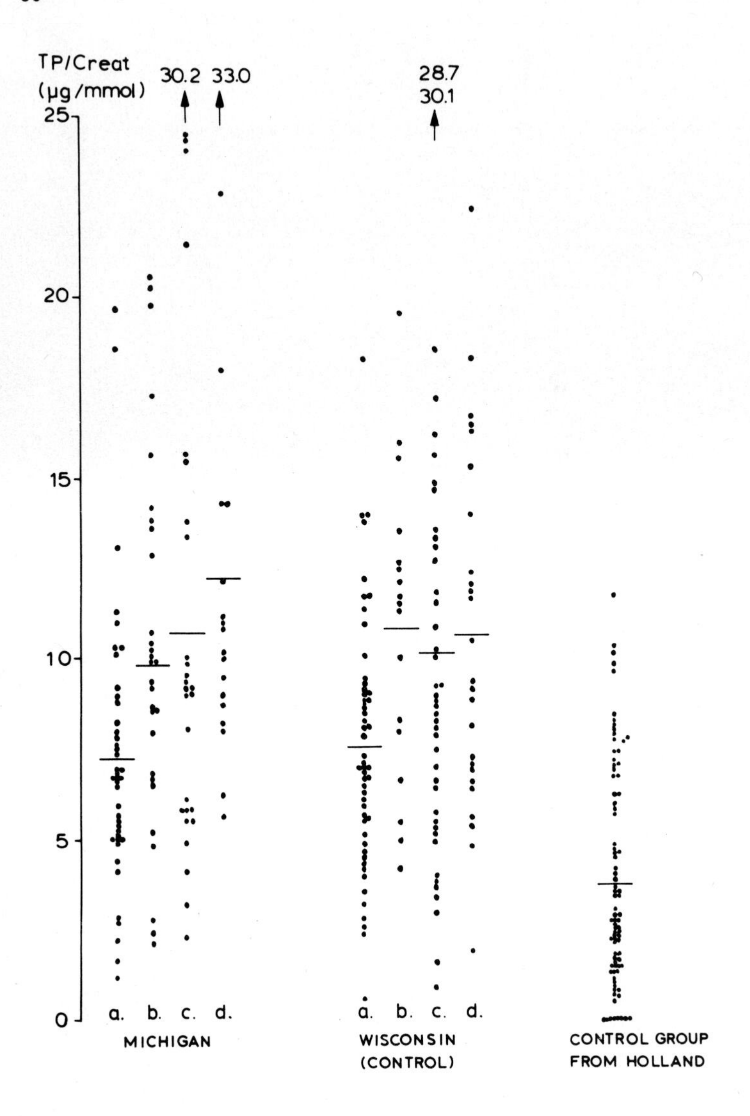

Fig. 2. Total porphyrin values (µg/mmol creatinine) in the Michigan and Wisconsin groups, subdivided in males, females, adults and young people and in a Dutch group.

	Michigan			Wisconsin		
	$\bar{x}$	s	n	$\bar{x}$	s	n
male , ≥ 19 years(a)	7.2	3.7	45	7.5	3.5	53
male , < 19 years(b)	9.8	5.3	31	10.8	4.2	17
female, ≥ 19 years(c)	10.7	7.2	31	10.1	7.1	42
female, < 19 years(d)	12.2	6.5	19	10.6	5.0	25

TABLE 5

TOTAL PORPHYRIN- (µg/l) AND CREATININE- (mmol/l) EXCRETION DURING 7 CONSECUTIVE DAYS IN NOVEMBER 1977 AND THOSE URINE VALUES FROM JUNE 1977 OF ONE MEMBER OF THE MICHIGAN GROUP

Day	Total porphyrins (T.P.) (µg/l)	Creatinine mmol/l	T.P./Creat. µg/mmol
1	76	4.8	15.9
2	97	13.7	7.1
3	105	11.7	8.9
4	80	3.6	22.2
5	105	8.2	12.8
6	142	11.9	12.0
7	79	7.0	11.3
Aver.	98	8.7	12.9
June '77	165	8.4	19.7

Thin-layer chromatographic separation of porphyrins

TLC plates were visually inspected and the relative amounts of the porphyrins present were recorded. The urines examined were placed in one of three categories on the basis of the porphyrins present. Table 6 shows that in 28 % of the Michigan samples examined, coproporphyrin was found in greater amounts than uroporphyrin, while, if present, traces of hepta-, hexa- and penta-carboxylic-porphyrin.

TABLE 6

DISTRIBUTION OF PORPHYRINS IN URINE OF PBB-EXPOSED FARM FAMILIES

Components Distribution	Number (n = 79)	% of n	% of total samples
c>u>>> (traces of 7,6,5)[a]	22	28	53[b]
c>u>5>7>6	11	14	9
c>u>7>5>6	46	58	38

[a] c = coproporphyrin; u = uroporphyrin; 7 = 7-COOH-porphyrin; 6 = 6-COOH-porphyrin; 5 = 5-COOH-porphyrin.

[b] 46 randomized samples were run for TLC. Additionally, 33 samples chosen by the high (>100 µg/l) total porphyrin value, were run for TLC as well. The samples left were considered to contain only traces of 7-, 6-, 5-COOH-porphyrins and are included in the 53 % figure.

In 14 % of the samples, a porphyrin pattern was found such that c>u>5>7>6, and in 58 % of the samples, a porphyrin pattern of c>u>7>5>6 was found.

In the Wisconsin group, those samples (37) were run for TLC which had a total porphyrin value above 11.0 µg/mmol creatinine. Also those samples (8), with a total porphyrin value below 3.0 µg/mmol creatinine were run for TLC. Another 8 samples between 3.0 and 11.0 µg/mmol creatinine (table 7) were taken at random.

TABLE 7

DISTRIBUTION OF PORPHYRINS IN THE WISCONSIN GROUP

Components distribution	Number (n=53)		
	<3.0	3.0<>11.0	>11.0[a]
c>u>>(traces of 7,6,5)	8	8	29
c>u>5>(traces of 7,6)	-	-	3
c>u>7>5>6	-	-	5

[a]µg/mmol creatinine

Table 7 shows that only in the total urinary porphyrin group exceeding > 11 µg/mmol creatinine in 5 out of 53 samples was an abnormal porphyrin pattern found.

The results of quantitative thin layer chromatography of the urinary porphyrins and porphyrin precursors in 15 samples randomly chosen from the Michigan group and in 4 not-exposed to PBB (laboratory workers) are compiled in tables 8 and 9.

ALA excretion was normal with one exception (case 12). Total porphyrins were elevated in more than half of the total studies (compared to European normal values). In detail, a secondary coproporphyrinuria was found in 4 cases, and a chronic hepatic porphyria Type A in 4 cases (table 10). In one case there was a transition between secondary coproporphyrinuria and chronic hepatic porphyria (case 17) (tables 8, 9 and 10). In two cases with secondary coproporphyrinuria (cases 1 and 16) as well as in one case with chronic hepatic porphyria Type A (case 16) we found a real decrease of the isomer III content of the urinary coproporphyrin, associated with an absolute increase of coproporphyrin isomer I excretion (table 8 and 9).

These observations correlate with the higer than normal amounts of bile pigments detected in TLC-films (table 8 and 9). Bilirubin and its derivatives are detectable in the TLC system used, when elevated in urine.

Both findings, the detection of elevated bile pigment excretion in urine by TLC and the decrease of the isomer III content of total coproporphyrin lead to the assumption that an intrahepatic cholestasis in the liver, as an expression of the PBB, caused hepatic injury (table 10).

In table 11 the relation between the porphyrin pattern and the corresponding average total porphyrin values is shown, expressed in µg/mmol creatinine. In preliminary experiments has been proven that the prophyrin excretion in morning urine is about half of that produced during 24 hours.

In the Michigan group there is no correlation between total porphyrin level and normal or abnormal porphyrin pattern. In the Wisconsin group an abnormal porphyrin pattern is found to be related to the highest urinary porphyrin level.

Resamplings

The results obtained from the two people who provided us with samples over consecutive days appear in table 12. The first person is male and 32 years old while the second is male and 16 years old. Though the porphyrin patterns found are classified under normal and Coproporphyrinuria, it is not true that all three conditions (c/u ratio, component and total porphyrins) are fulfilled. It is sometimes difficult to see a clear difference between the chronic hepatic porphyrias, especially between Coproporphyrinuria and CHP Type A_1. In this study the relative distribution of the components has been chosen to be the major condition needed to classify a porphyrin pattern. Therefore tabel 13 has been developed.

The different values found for the 27 resampled people are given in tabel 14. No relation between porphyrin pattern and total porphyrin excretion was found. The number of abnormal porphyrin patterns found in the samples from June seemed to be less than those found in November.

Urinary porphyrins of relatives of people from Michigan living outside the state

The average amounts of total porphyrin (µg/l and µg/mmol creatinine) and creatinine (mmol/l) in the morning urine of members of this group are given in tabel 15. There are no significant differences from those values found with the exposed people from Michigan (figs. 1 and 2). Porphyrin patterns have not been examined.

TABLE 8

URINARY PORPHYRINS AND PRECURSORS IN PBB-EXPOSED FARM FAMILIES

Case no.	sex	age (years)	ALA	PBG	PORPHYRINS								Bile pigments in TLC
					total	U	7	6	5	C	3	CIII	
			mg/l		µg/l	%						% isomer	
1	F	35	2.8	1.1	146 ↑	9	2	1	2	82	4	67 ↓	↑
2	F	29	2.4	0.5	3	14	8	3	3	69	3	73	
3	M	47	2.5	0.6	77	12	3	1	3	78	3	78	(↑)
4	F	24	2.8	0.5	41	13	7	2	2	74	2	72	
5	M	50	2.7	0.5	61	23	3	2	3	67	2	74	↑
6*	M	30	2.8	0.5	64	16	2	1	3	76	2	71	
7	M	39	3.2	0.5	123 ↑	10	5	1	2	80	2	75	
8*	M	34	2.9	0.4	87	16	2	1	3	76	2	70	↑
9	M	32	4.7	0.5	171 ↑	16	4	1	2	75	2	78	↑
10	M	47	3.5	0.7	117 ↑	12	2	1	3	80	2	75	
11*	F	29	2.9	0.5	68	16	7	1	6	69	1	73	↑
12	M	35	6.8	0.6	272 ↑	16	6	1	3	73	1	84 (↑)	↑↑
13	M	11	3.0	0.4	87	12	7	1	2	76	2	76	
14	F	< 19	3.4	0.5	115 ↑	11	2	1	3	80	3	67 ↓	↑
15	M	31	2.9	0.4	59	8	2	2	5	80	3	69	↑↑
16	F	> 19	6.9	0.6	309 ↑	18	6	1	3	70	2	67 ↓	↑
17	M	36	2.6	0.6	173 ↑	15	3	1	4	75	2	75	
18*	F	32	2.5	0.4	39	23	2	2	2	69	2	73	
19	F	14	5.3	0.5	215 ↑	15	7	1	3	73	1	70	↑
Controls (n = 1360)			2.9 ± 1.6 ($\bar{x} \pm s$)	0.8 ± 0.4 ($\bar{x} \pm s$)	61 ± 17 ($\bar{x} \pm s$)							(69-83)	
Upper limit of normal			6.5	1.6		21	3	2	4	69	1		

Abbreviations: ALA δ-aminovulinic acid; PBG porphobilinogen; U uroporphyrin; 7 heptacarboxylicporphyrin; 6 hexacarboxylicporphyrin; 5 pentacarboxylicporphyrin; C coproporphyrin; 3 tricarboxylicporphyrin; TLC thin layer chromatography; M male; F female. * control ↑↓in-/decreased

TABLE 9

URINARY PORPHYRINS AND PRECURSORS IN PBB-EXPOSED FARM FAMILIES

Case no.	sex	age (years)	ALA	PBG	PORPHYRINS total	U	7	6	5	C	3	CIII	Bile pigments in TLC
			mg/l	mg/l	µg/l	µg/l	µg/l	µg/l	µg/l	µg/l	µg/l	% isomer	
1	F	35	2.8	1.1	146 ↑	13	3	<1	3	120	6	67 ↓	↑
2	F	29	2.4	0.5	3	5	3	1	1	24	<1	73	
3	M	47	2.5	0.6	77	10	2	1	2	60	2	78	(↑)
4	F	24	2.8	0.5	41	5	3	1	1	30	<1	72	
5	M	50	2.7	0.5	61	14	2	1	2	41	1	74	↑
6*	M	30	2.8	0.5	64	10	1	<1	2	49	1	71	
7	M	39	3.2	0.5	123 ↑	12	6	1	3	98	3	75	
8*	M	34	2.9	0.4	87	13	2	1	3	66	2	70	↑
9	M	32	4.7	0.5	171 ↑	29	6	1	4	128	3	78	↑
10	M	47	3.5	0.7	117 ↑	14	2	1	4	95	2	75	
11*	F	29	2.9	0.5	68	11	5	1	4	47	1	73	↑
12	M	35	6.8	0.6	272 ↑	44	15	3	8	199	3	84 (↑)	↑↑
13	M	11	3.0	0.4	87	10	6	1	2	66	2	76	
14	F	<19	3.4	0.5	115 ↑	13	2	<1	3	93	3	67 ↓	↑
15	M	31	2.9	0.4	59	5	1	<1	3	47	2	69	↑↑
16	F	>19	6.9	0.6	309 ↑	59	18	2	8	218	4	67 ↓	↑
17	M	36	2.6	0.6	173 ↑	26	5	2	7	130	3	75	
18*	F	32	2.5	0.4	39	9	1	<1	1	27	<1	73	
19	F	14	5.3	0.5	215 ↑	32	15	2	6	158	2	70	↑
Controls (n = 1360)			2.9±1.6 ($\bar{x} \pm s$)	0.8±0.4 ($\bar{x} \pm s$)	61±17 ($\bar{x} \pm s$)							(69-83)	
Upper limit of normal			6.5	1.6		24	3	2	4	78	1		

Abbreviations: ALA δ-aminovulinic acid; PBG porphobilinogen; U uroporphyrin; 7 heptacarboxylicporphyrin; 6 hexacarboxylicporphyrin; 5 pentacarboxylicporphyrin; C coproporphyrin; 3 tricarboxylicporphyrin; TLC thin layer chromatography; M male; F female * control ↑↓in-/decrease

TABLE 10

DIAGNOSIS OF DISTURBANCE OF HEPATIC PORPHYRIN METABOLISM IN PBB-EXPOSED FARM FAMILIES

Case (see table 8 and 9)	Diagnosis
1	Secondary coproporphyrinuria with intrahepatic cholestasis
7	Secondary coproporphyrinuria
9	Chronic hepatic porphyria Type A
10	Secondary coproporphyrinuria
12	Chronic hepatic porphyria Type A
14	Secondary coproporphyrinuria with intrahepatic cholestasis
16	Chronic hepatic porphyria Type A with intrahepatic cholestasis
17	Secondary coproporphyrinuria with transission to CHP A
19	Chronic hepatic porphyria Type A

TABLE 11

RELATION BETWEEN PORPHYRIN PATTERN AND TOTAL PORPHYRIN VALUES (μg/mmol creatinine) IN THE EXAMINED SAMPLES

	Michigan			Wisconsin		
	$\bar{x}$	s	n	$\bar{x}$	s	n
c>u>>(traces of 7,6,5)	10.3	8.6	22	11.1[a]	7.0	45
c>u>5>(traces of 7,6)	10.0	7.3	11	13.2	1.5	3
c>u>7>5>6	9.0	4.2	46	19.4[b]	6.0	5

[b]Significantly elevated towards[a] (Wilcoxon test, P=0.05)

TABLE 12

TOTAL PORPHYRIN VALUES (μg/l and μg/mmol creat.), CREATININE VALUE (mmol/l) AND PORPHYRIN PATTERN OF THE TWO PEOPLE WHO PROVIDED SAMPLES DURING CONSECUTIVE DAYS

Day of sampling	Total porphyrins (μg/l)	Total porphyrins (μg/mmol creat.)	Creatinine (mmol/l)	Porphyrin pattern
12/11[a]	76	15.9	4.8	Normal
13/11	97	7.1	13.7	Normal
14/11	105	8.9	11.7	Normal
15/11	80	22.2	3.6	
16/11	105	12.8	8.2	
17/11	142	12.0	11.9	
18/11	79	11.3	7.0	
Average	98(s=23)	12.9(s=5.0)	8.7(s=3.8)	
June, 1977	165	19.7	8.4	Normal
16/11[b]	99	7.5	13.3	Normal
19/11	89	6.0	14.9	Normal
20/11	91	4.9	18.5	Normal
21/11	127	6.3	20.1	
Average	101(s=18)	6.2(s=1.1)	16.7(s=3.1)	
June, 1977	41	2.1	19.8	Coproporphyrinuria

[a]Male, 32 years old [b]Male, 16 years old

TABLE 13

APPEARANCE OF DIFFERENT PORPHYRINS IN URINE DETECTED BY TLC AND THEIR CLASSIFICATION

x-COOH-porphyrins					Classification
copro(4)	penta(5)	hexa(6)	hepta(7)	uro(8)	
+	-	-	-	+	
+	t	-	-	+	c>u>>(traces of
+	-	-	t	+	7,6,5):
+	t	-	t	+	Normal
+	t	t	t	+	
+	+	-	-	+	
+	+	-	t	+	c>u>5>7>6 :
+	+	t	t	+	Coproporphyrinuria
+	+	-	+	+	and CHP
+	+	t	+	+	Type A_1
+	-	-	+	+	
+	t	-	+	+	
+	+	-	+	+	
+	-	t	+	+	c>u>7>5>6 :
+	t	t	+	+	CHP Type A_2
+	+	t	+	+	
+	+	+	+	+	

+ = present and measurable on the Beckman Spectrofotometer Acta C III, Extinction > 0.009 at 404 nm. 1 cm cuvet

- = not present

t = present and not measurable on the Beckman Spectrofotometer Acta C III

TABLE 14

TOTAL PORPHYRIN VALUES (μg/l and μg/mmol creatinine), CREATININE VALUE (mmol/l) AND PORPHYRIN PATTERN IN THE URINE OF 27 RE SAMPLED PEOPLE FROM MICHIGAN

Sex - Age	Total porphyrins (μg/l)[a]		Total porphyrins μg/mmol creat.)		Creatinine mmol/l)		Porphyrin pattern[b]	
	June	Nov.	June	Nov.	June	Nov.	June	Nov.
F - 33	44	5	24.5	0.8	1.8	6.4	-	-
M - 33	-	60	-	4.8	-	12.3	-	-
M - 11	94	83	-	6.9	-	11.9	-	-
F - 9	130	30	11.1	4.8	11.7	6.2	-	-
M - 38	132	105	13.1	5.8	10.1	18.1	A_2	A_1
M - 10	182	71	10.4	7.8	17.5	9.1	A_2	N
F - -								
F - 7	74	69	14.3	14.9	5.2	4.6	-	-
F - 64	88	27	24.4	9.5	3.6	2.8	A_2	N
M - 35	132	175	6.7	8.8	19.7	19.9	A_1	A_2
M - 63	83	38	10.3	4.5	8.0	8.4	-	-
F - 63	50	29	15.5	4.9	3.2	6.0	N	-
M - 34	105	247	4.9	9.4	21.5	26.3	A_2	A_2
F - 32	72	180	13.8	11.0	5.2	16.3	-	-
M - 56	96	37	6.7	5.8	14.6	6.4	C	A_2
F - 54	66	15	15.7	3.5	4.2	4.2	-	-
M - 28	-	83	-	11.8	-	7.0	-	-
M - 25	44	60	2.2	4.5	20.1	13.3	-	-
M - 26	121	87	8.2	8.6	14.7	10.1	A_1	N
M - 18	-	68	-	7.2	-	9.5	-	-
M - 51	110	127	6.9	7.4	15.9	17.1	A_2	A_1
F - 46	66	39	9.4	6.0	7.0	6.6	-	-
M - 43	138	35	18.6	10.8	7.4	3.2	A_1	N
F - 40	-	40	-	1.3	-	29.6	-	-
M - 16	88	76	10.2	4.5	8.6	16.9	A_2	A_2
M - 20	-	85	-	8.1	-	10.5	-	-
M - 21	-	121	-	12.3	-	9.9	-	-

TABLE 15

TOTAL PORPHYRIN VALUES (μg/l and μg/mmol creat.) AND CREATININE VALUE (mmol/l) IN THE CONTROL GROUP, FORMED BY RELATIVES LIVING OUTSIDE THE STATE OF MICHIGAN

	$\bar{x}$	s	n
Total porphyrins (μg/l)	83	62	32
Total porphyrins (μg/mmol creatinine)	9.0	5.8	32
Creatinine (mmol/l)	10.6	5.9	32

Serum enzyme activities and urinary porphyrin data of PBB exposed individuals

In table 16 the average enzyme activities (SGOT, SGPT, LDH) and the PBB-levels are given, corresponding to the four recognized porphyrin patterns (Normal, Coproporphyrinuria, CHP Type A_1 and CHP Type A_2). No correlation is found between serum enzyme activities, porphyrin pattern and PBB levels. A high PBB serum level seems to correlate with an abnormal porphyrin pattern.

TABLE 16

SGOT-, SGPT-, LDH-VALUES (U/l) AND PBB-VALUES (pbb) IN BLOOD, CORRESPONDING TO THE FOUR PORPHYRIN PATTERNS FOUND IN 46 PEOPLE OF THE MICHIGAN GROUP

	N[a]	C	A_1	A_2
n	12	5	11	18
SGOT	25.3± 8.9	25.0± 6.2	27.5± 9.8	20.3±11.8
SGPT	37.0±26.6	30.0±21.4	20.1±13.6	19.8±18.3
LDH	192 ±96	175 ±60	195±45	151 ±42
n	11	5	9	10
PBB	3.1± 3.4	4.9± 7.2	5.0± 9.1	4.9± 9.7

[a]N = Normal
C = Coproporphyrinuria
A_2= CHP Type A_2

Figure 3 gives the distribution of the SGOT-, SGPT- and LDH-values over the four porphyrin patterns. In this figure the average values and the normal ranges of values are also given.

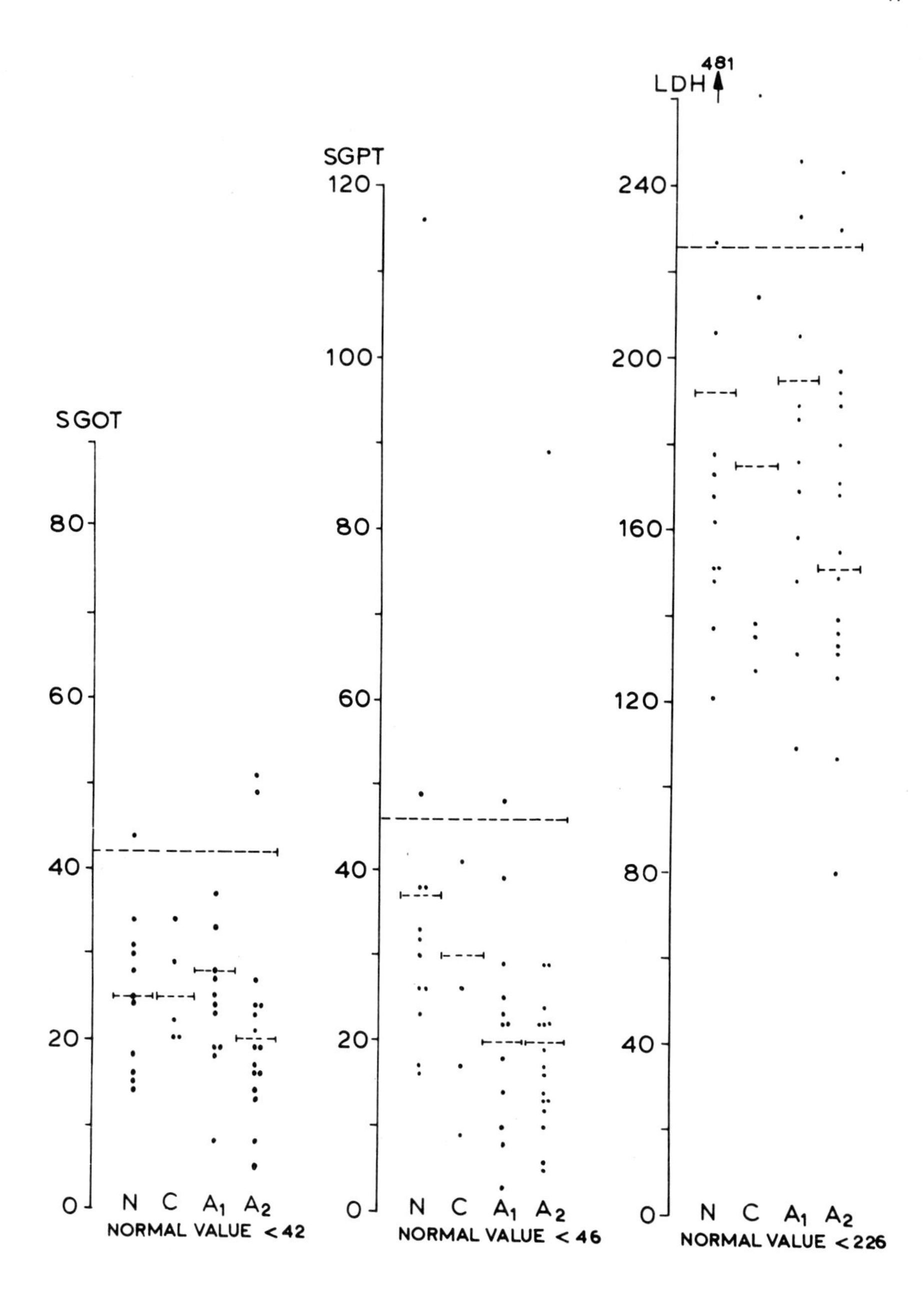

Fig. 3. Distribution of the SGOT-, SGPT-, and LDH-values over the four porphyrin patterns of 46 people of the exposed Michigan group.

DISCUSSION

Polyhalogenated hydrocarbons (HCB, 1,4 dichlorobenzene, 1,2,4-trichlorobenzene, octachlorostyrene, PCB, hexachlorcbiphenyl isomers, TCDD, PBB, γ-hexachlorocyclohexane, methylchloride and vinylchloride) have been shown to be porphyrinogenic in experimental animals, and, in some cases, in man[12]. A feature of the porphyria evoked by these halogenated aromatics is the slow onset. A chronic exposure is usually needed, especially in mammals[13]. Small birds, e.g., Japanese quail, are often used in laboratory experiments because of their sensitivity to the action of certain halogenated aromatic compounds and the rapidity with which they develop porphyria. Compared to the porphyrinogenic action of HCB and PCB, PBB is respectively 80 and 10 times more active in a Japanese quail[14,15].

Hexachlorobenzene induced porphyria in the rat provides a suitable experimental model for studying the stages of development of human chronic hepatic porphyria[16]. Just as with chronic hepatic porphyria in man, the development of experimental hexachlorobenzene porphyria in the rat can be divided into several stages, chronic hepatic porphyria Type A, B, C and D. In hexachlorobenzene porphyria in rats chronic hepatic porphyria Type A develops after a relatively long interval of secondary coproporphyrinuria. The primary enzymatic defect in hexachlorobenzene porphyria in rats[16] as well as in the chronic hepatic porphyrias in man is the diminished activity of uroporphyrinogen decarboxylase in the liver[17]. The increase of uro- and heptacarboxylicporphyrin excretion in the urine and the accumulation of both these porphyrins in the liver reflect directly the specific enzymatic defect in experimental and human chronic hepatic porphyria.

The chronic hepatic porphyrias are the most common hepatic porphyrias in humans in Europe[18], and they are associated with various and different hepatic lesions, such as fibrosis, chronic aggressive hepatitis, fatty liver, siderosis and cirrhosis. For example, all stages of chronic hepatic porphyria, from Type A up to Type D, are found in liver cirrhosis of various pathogeneses[19].

As a group, polyhalogenated aromatics are very stable, persistent compounds. They resist degradation and accumulate in the food chain, concentrating ultimately in the fatty tissues of fish-eating birds[20] and man. Their long term action results in liver cell damage. One of the first measurable signals for this liver damage is the increase of porphyrins with 8 and 7 carboxylic groups in the liver and urine.

For the purpose of this study, we have identified a group which has been exposed to a powerful porphyrinogen, over the past several years. The sole purpose of this investigation was to assess liver health, using urine porphyrin excretion as an indicator, in a population of farm families, which we presume, received the highest doses of PBB related to a proper control group from Wisconsin with no known exposure to PBB. These two groups are composed of persons who have similar life styles and working habits and are predominantly members of dairy farm families. No attempt has been made to analyze or assess any other PBB produced conditions, symptoms, or illnesses, unrelated to liver damage.

Total urinary porphyrins from Michigan farm families exposed to PBB did not differ from values from the Wisconsin control group, whether expressed in μg/l nor in μg/mmol creatinine.

When compared to data from a Dutch control group and to data from Doss, the American values can be seen to be increased above the European values. The same was true concerning creatinine excretion of the European and American groups. Diet or genetic factors play a possible role.

The results presented in table 12 concerning a person from Michigan whose urine was sampled in June and November (during seven consecutive days), give support to the conclusion that in the range up to 200 μg/l of total porphyrins a conclusion concerning liver damage cannot be reached just by looking at total porphyrins. Although certain values exceed 100 μg/l, according to European values >100 μg/l is thought to be abnormal[10,18].

Just looking at the total porphyrin excretion in the range up to 200 μg/l does not give an indication of exposure to PBB (this study).

In the Michigan group it has been found that at least 47 % of the people have a porphyrin pattern which is not normal (table 6). According to Doss,[10,18] we can classify the 47 % under Coproporphyrinuria, CHP Type A_1 and CHP Type A_2. However, it is difficult to specify the samples to one of these, because one or more of the conditions (c/u ratio, porphyrin pattern and total porphyrins) did not always agree in those samples.

Of the 53 examined samples of the Wisconsin group, eight have been found to be abnormal. These eight samples all have a total porphyrin amount above 11.0 μg/mmol creatinine (table 7). The 84 samples, which have not been examined, all have a total porphyrin amount between 3.0 and 11.0 μg/mmol creatinine. If it is assumed that no abnormals will be found in this group, it means that only 6 % of the people in the Wisconsin group have an abnormal porphyrin pattern; these can also be classified under Coproporphyrinuria, CHP Type A_1 and CHP Type A_2.

This assumption is supported by the observation that the total urinary porphyrin excretion in the Wisconsin group is higher in the subgroup with an abnormal porphyrin pattern. In the Michigan group there is no correlation between urinary total porphyrin excretion and porphyrin pattern. The change in the porphyrin pattern indicates a porphyrinogenic factor has to be involved. Exposure to the porphyrinogenic compound PBB might be the causative factor.

Resamplings

The results in table 12 of total porphyrins and porphyrin pattern confirm the previous conclusion: "The amount of porphyrins, in the range up to 200 μg/l, does not give an indication of exposure to PBB".

For the first person a normal porphyrin pattern is found in the case of excretion of 76, 97 and 105 μg/l, also when the excretion is 165 μg/l.

The second person has a normal pattern when there is an excretion of 99, 89 and 91 μg/l, while when Coproporphyrinuria is found excretion is 41 μg/l. In the 5 months between the sampling data a pattern which is characteristic for Coproporphyrinuria has changed into a pattern which can be called Normal. This means that a disturbance in porphyrin excretion is reversible and can disappear again.

From table 14 it can also be concluded that the porphyrin pattern can change in time. With the examined samples it can be seen that some pattern return from Abnormal to Normal; the opposite also happens (CHP Type A_1 and Coproporphyrinuria change into CHP Type A_2).

Because the porphyrin pattern can, in time, return to normal in this case a most severe CHP Type A_2 goes back to normal, the right moment of sampling is very important.

Urinary porphyrins of relatives of examined people from Michigan, who live outside the state

Almost all examined people have been with their relatives for short or longer stays, during the period 1973 - 1977. It is very likely that they were exposed to PBB during those stays. How long and how much is not known. Total porphyrin values of these people show no significant differences from the values of their relatives from Michigan, as we might expect after our previous conclusion.

It is also advisable that the porphyrin pattern of these people should be determined, before we can make further conclusions.

Serum enzyme activities and urinary porphyrins from PBB exposed individuals

From the above results it may be concluded that, though there are some values of SGOT, SGPT and LDH above the normal range values, there is no correlation between these values and the porphyrin pattern. This means that at the time of sampling, people who have an abnormal porphyrin pattern do not show a liver damage which causes a liberation of these enzymes into the blood.

There is also no correlation between porphyrin pattern and PBB-values found in blood. This is to be expected, because in earlier studies[21,22] it was found that there was no correlation between PBB serum levels and symptoms of illnesses and complaints. In these studies it was also found that there was no constant relationship between fat levels of PBB and serum levels. After exposure of individiuals to PBB, the PBB serum levels will taper off to very low or near zero levels. Because PBB is highly fat soluble, it would be much better to correlate symptoms with fat levels of PBB. This is currently being undertaken in the study by Selikoff and co-workers[22].

An average urinary total porphyrin value of 77 μg/l was found (this volume) with the people from Seveso living in an area contaminated by TCDD, 84 % of the people had a prophyrin pattern which was abnormal, the most developed being CHP Type A_2.

For the Yusho patients exposed to PCB an average urinary total porphyrin value of 79 μg/l was found (this volume). One of the patients and one of the controls had an abnormal porphyrin pattern (CHP, Type A_2).

From these values the conclusion is made: total urinary porphyrin values in the range up to 200 μg/l are no indication for exposure to polyhalogenated aromatics; the pattern of porphyrins is.

In analogy to the toxic role of alcohol which can also produce a secondary coproporphyrinuria, some times associated with a δ-aminolevulinaciduria, PBB can produce similar abnormalities in porphyrin excretion perhaps on the basis of a direct disturbance of enzymes of the heme biosynthetic pathway, coproporphyrinogenoxidase and δ-aminolevulinic acid dehydratase, with or without a morphological lesion of the liver cells[17] and induction of δ-aminolevulinic acid synthase[12,23,24]. In a second stage of intoxication - as in the development of hexachlorobenzene chronic hepatic porphyria in rats[25] - the uroporphyrinogen decarboxylase in the liver is affected by interference of mixed function oxidase metabolism of polyhalogenated aromatics[12,13] and its activity decreases resulting in the biochemical manifestation of chronic hepatic porphyria Type A[25]. In man without features of intoxication with hexachlorobenzene, PBB or PCB, an inherited defect of uroporphyrinogen decarboxylase in combination with a liver injury[25] is probably sufficient pathological background for the gradual development of chronic hepatic porphyria.

In conclusion, hepatic uroporphyrinogen decarboxylase can be disturbed genetically as well as by an intoxication, e.g. by hexachlorobenzene, PBB or PCB.

In the PBB-exposed persons a prolonged coproporphyrinuria precedes the chronic hepatic porphyria Type A.

ACKNOWLEDGEMENTS

The authors would like to express their appreciation to the doctoral students Els van Vliet and Helmi Wetzels who improved our urinary porphyrin analysis and ran a control group, to Carol Robertson for expert technical assistance, to Ans Strik whose moral support and patience made this project possible, to Prof.Dr. K. Biersteker and Drs. T. Taselaar (Dept. of Tropical Health and Hygiene, Agricultural University) for their help in the design of the group selections, Barbara Lear, P.A. (Marshfield Medical Foundation, Wisconsin, U.S.A.) for her help in collecting urine samples from Wisconsin, and to Prof.Dr. J. Selikoff and Dr. H. Anderson (Environmental Health Sciences Laboratory, Mount Sinai School of Medicine, City University of New York) for providing clinical chemistry data.

Part of this study (M.D.) was supported by the Deutsche Forschungsgemeinschaft (grant Do 134/7).

REFERENCES

1. Cam, C. and Nigogosyan, G. (1963) J.Amer.Med.Assoc., 183, 88.
2. Bleiberg, J., Wallen, M., Brodkin, R., Applebaum, J.L. (1964) Arch.Derm., 89, 793.
3. Jirásek, L., Kalenský, J., Kubec, K., Pazderová, J. and Lukás, E. (1976) Hautartz, 27, 328.
4. Poland, A., Smith, D., Metter, C., Possick, P. (1971) Arch.Environ. Health, 22, 316.
5. Chalmers, J.N.M., Gillam, A.E. and Kench, J.E. (1940) Lancet, 28 dec., 806.
6. Lange, C.E., Block, H., Veltman, G. and Doss, M. (1976) Urinary Porphyrins Among PVC Workers, in M. Doss (Ed.), Porphyrins in Human Diseases, Karger, Basel, p. 352.
7. Robertson, W. and Chynoweth, P. (1975) Environment, 17, 25.
8. Wolff, M.S., Aubrey, B., Camper, F. and Haymes, M. (1978) Environm.Health Perspect, 23, 177.
9. Strik, J.J.T.W.A. (1973) Meded. Rijksfac. Landbouwwet. Gent, 38, 709.
10. Doss, M. (1974) Porphyrins and Porphyrin Precursors, in M.C. Curtius and M. Roth (Eds.), Clinical Biochemistry, Principles and Methods, De Gruyter, Berlin, Vol. II, p. 1339.
11. Gorter, E. and Graaff, W.C. de (1955) Klinische Diagnostiek, Stenfert Kroese N.V., Leiden, 7th ed., p. 440.
12. Strik, J.J.T.W.A. (1977) Porphyrinogenic Action of Polyhalogenated Aromatic Compounds with special Reference to Porphyria and Environmental Impact, in M. Doss (ed.), Proc.Int.Symp.Clin.Biochem. Diagnosis and Therapy of Porphyrias and Lead Intoxication, Springer-Verlag, Berlin, p. 151.

13. Strik, J.J.T.W.A. (1973) Enzyme, 16, 224.

14. Strik, J.J.T.W.A. (1973) Experimental hepatic porphyria in birds, Thesis, Utrecht.

15. Vos, J.G., Strik, J.J.T.W.A., Holsteijn, C.W.M. and Pennings, J.H. (1971) Toxicol. Appl. Pharmacol., 20, 232.

16. Doss, M., Schermuly, E. and Koss, G. (1976) Ann. Clin. Res., 8 (suppl. 17), 171.

17. Doss, M., Schermuly, E., Look, D. and Henning, H. (1976) Enzymatic Defects in Chronic Hepatic Porphyrias, in M. Doss (Ed.), Porphyrins in Human Diseases, Karger, Basel, p. 286.

18. Doss, M., Look, D., Henning, H., Lüders, C.J., Dölle, W. and Strohmeijer, G. (1971) Z. Klin. Chem. u. Klin. Biochem., 9, 471.

19. Doss, M., Look, D., Henning, H., Nawrocki, P., Schmidt, A., Dölle, W., Korb, G., Lüders, C.J. and Strohmeijer, G. (1972) Klin. Wschr., 50, 1025.

20. Koeman, J.H., Velzen-Blad, H.C.W. van, Vries, R. de and Vos, J.G. (1973) J. Reprod. Fertil., Suppl. 19, 353.

21. Meester, W.D. and McCoy, D.J. (1976) Human Toxicology of polybrominated biphenyls. Symposium on Environmental Toxicology, Seattle, Washington, Aug. 4.

22. Selikoff, J. (1977) Health Effects of Exposure to Polybrominated Biphenyls. Results of a Clinical Field Survey, November, 4, 1976. Interim Summary Report to the National Institute of Environmental Health Sciences E.S. 00928.

23. Strik, J.J.T.W.A. (1973) Enzyme, 16, 211.

24. Strik, J.J.T.W.A. and Koeman, J.H. (1976) Porphyrinogenic Action of Hexachlorobenzene and Octachlorostyrene, in M. Doss (ed.), Porphyrins in Human Diseases, Karger, Basel, p. 48.

25. Doss, M. (in press) Klin. W.schr.

Chemical Porphyria in Man, J.J.T.W.A. Strik and J.H. Koeman eds.

PORPHYRINS AS POSSIBLE PARAMETERS FOR EXPOSURE TO HEXACHLOROCYCLOPENTADIENE, ALLYLCHLORIDE, EPICHLOROHYDRIN AND ENDRIN

A. NAGELSMIT[a], P.W. VAN VLIET[a], W.A.M. VAN DER WIEL-WETZELS[a], M.J. WIELARD[a], J.J.T.W.A. STRIK[a], C.F. OTTEVANGER[b] AND N.J. VAN SITTERT[c]

[a]Department of Toxicology, Agricultural Univers'ty, Wageningen, The Netherlands

[b]Department of Occupational Health, Shell Nederland Chemie, Rotterdam, The Netherlands

[c]Shell Internationale Research Maatschappij, The Hague, The Netherlands

SUMMARY

Japanese quail received a daily dose of 200 mg/kg allylchloride, dissolved in oil. After a dosing period of 15 days macroscopic and microscopic fluorescence of porphyrins in the intestines and their contents were observed. No fluorescence was found in liver and kidneys. The intestines of HCB (hexachlorobenzene) treated quail (positive controls) showed a strong fluorescence and the liver and kidneys a very weak fluorescence. No fluorescence was seen in tissues of qual treated with hexachlorocyclopentadiene, epichlorohydrin and endrin.

The urinary porphyrin excretion was studied in groups of industrial workers. None of the groups of workers, exposed to either allylchloride, hexachlorocyclopentadiene, epichlorohydrin or endrin showed an increase in urinary total porphyrin level as compared to a control group of office employees. The porphyrin patterns of the exposed groups were also normal.

INTRODUCTION

Several compounds are known to interfere with heme synthesis in experimental animals and cause porphyrin accumulation in organs and increase porphyrin excretion in urine[5]. Some chlorinated hydrocarbons such as vinylchloride[6], methylchloride[7] and hexachlorobenzene[8] have proved to have porphyrinogenic properties in humans too. The possible porphyrinogenic action pf hexachlorocyclopentadiene (HCCP), allylchloride (AC), epichlorohydrin (ECH) and endrin was tested by dosing the compounds orally in capsules to Japanese quail. Japanese quail were used because of their high sensitivity with respect to the porphyrinogenic action of chemicals[5].

The following compounds were studied

(a) Hexachlorocyclopentadiene (HCCP):

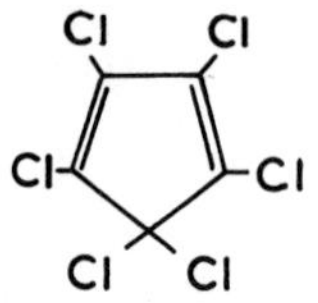

HCCP is an intermediate in the manifacturing of the insecticides aldrin, dieldrin and endrin.

(b) Allylchloride (AC):

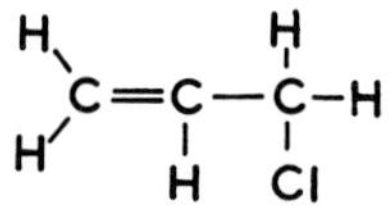

AC is an intermediate in the manifacturing of epichlorohydrine

(c) Epichlorohydrin (ECH):

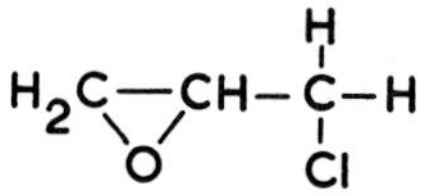

ECH is an intermediate in the manifacturing of glycerine

(d) Endrin:

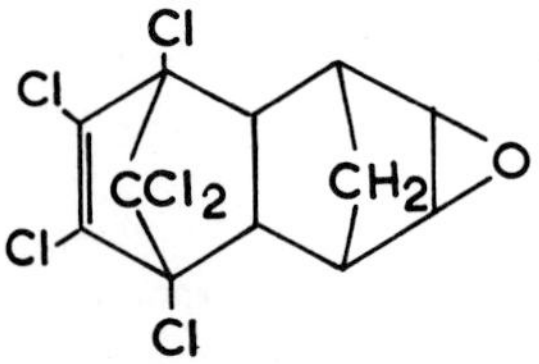

Endrin is used as a pesticide for agricultural purposes.

The porphyrinogenic properties of the compounds described have been studied in Japanese quail by means of fluorescence techniques[5]. Longterm exposure in man to porphyrinogenic compounds e.g. TCDD, HCB, methyl- and vinylchloride, results in an increase in the amount of porphyrins in urine as well as in an abnormal porphyrin pattern[11]. For this reason urines of male workers, producing the mentioned compounds, were analysed.

MATERIALS AND METHODS

Testing in experimental animals of porphyrinogenic properties of compounds used

The animals were acclimatized to the experimental conditions for a period of 3 days. Groups of 2-6 were formed at random. The quail received one gelatin capsule per day, administered orally, containing one of the compounds (HCCP, AC, ECH, Endrin) dissolved in olive oil. Since porphyria in quail generally develops within a few weeks exposure to high dose levels, an attempt was made to administer the compounds at the maximum level tolerated during the test period.

The following dose levels were applied:

HCCP (♀♀+♂♂)	100-200-300 mg/kg
AC (♀♀+♂♂)	50-100-200 mg/kg
ECH (♀♀)	50-100-200-300 mg/kg
Endrin (♀♀)	1-3-5- mg/kg

Initially AC was administered in dosages of 600-400-200 mg/kg respectively. As high mortality occurred, the dosages were largely reduced. Due to high mortality the experiment in the group exposed to 5 mg endrin/kg was discontinued.

The negative controls daily received a gelatin capsule with olive oil. One group of quail, which served as a positive control, received a daily dose of 1000 mg/kg HCB (in olive oil by gelatin capsule).

No food was given during the test period. After dissection fluorescence of the porphyrins was detected macroscopically as well as microscopically[5]. Special attention was focused on the liver, kidneys and intestines.

Urine analysis

Morning urine samples of the workers were analysed according to Doss[12,13] and this volume. The number of persons in the groups are given in Table 1.

TABLE 1
NUMBER OF WORKERS IN DIFFERENT PLANTS FROM WHICH URINARY PORPHYRINS WERE ANALYSED

Compound	Number of individuals per group	Number of individuals sampled twice
HCCP	40	35
AC	61	44
ECH	28	-
Endrin	39	-
Controls	24	-

Morning urine samples obtained from office workers of the same chemical company served as a control group. In the groups exposed to HCCP and AC urine was sampled twice. The porphyrins were separated by T.L.C. according to Doss[13] and the pattern was judged under U.V.-light.

RESULTS

Testing in experimental animals of porphyrinogenic properties of compounds used

No fluorescence was observed in tissues of quail exposed to HCCP, ECH, endrin resp. The exposure to 200 mg/kg AC during a period of 15 days resulted in a clear fluorescence of the intestines and their contents, whereas exposure to lower concentrations gave no visible fluorescence. The positive controls exposed to HCB showed fluorescence of the intestines and a very weak fluorescence of the liver and the kidneys. No liver fluorescence at all was obtained in the livers of animals exposed to AC. The group serving as a negative control did not show any fluorescence.

Urine analysis

The total mean amounts of urinary porphyrins (µg/l) of plant workers and controls are given in Table 2.

TABLE 2

TOTAL MEAN URINARY PORPHYRIN VALUES OF PLANT WORKERS AND CONTROLS

Group	Number of individuals	Mean	S.D.	S.E.M.
HCCP 1	40	71.7	60.7	9.6
2	35	60.2	46.3	7.8
AC 1	61	51.9	33.3	4.3
2	44	41.1	32.8	4.9
ECH	28	72.1	60.7	11.5
Endrin	39	46.1	54.2	8.5
Controls	23[1]	63.4	41.0	8.6

[1]Because of an extremely high value (460 µg/l) one of the 24 controls was excluded from the statistics

No significant difference was found between exposed groups and the control group by student-t-test ($\gamma = 0.05$). The individual values of porphyrin amount are plotted in the scatter diagram (Fig. 1).

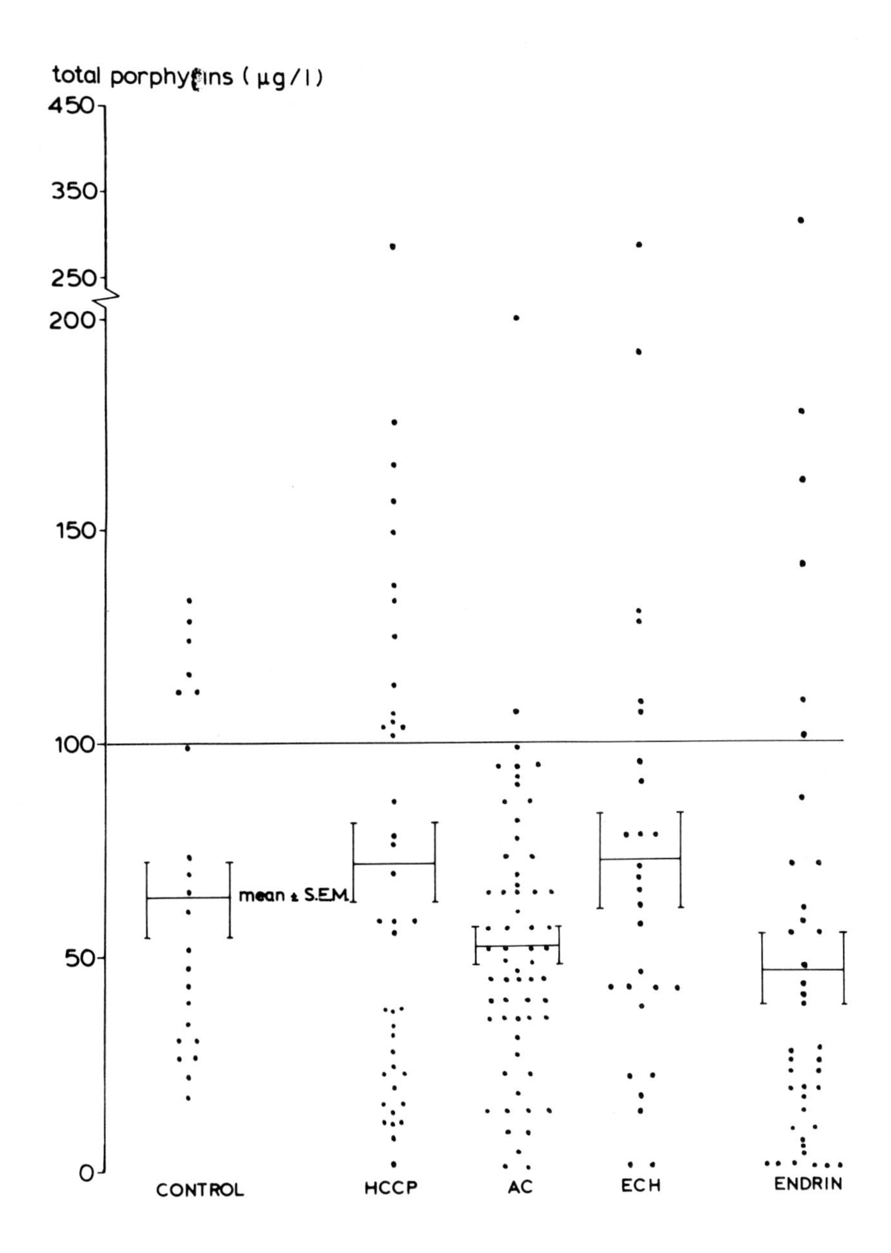

Fig. 1. Total urinary porphyrin values of plant workers and controls. For abbreviations used see Introduction.

Only the results of the first sampling in the groups exposed to HCCP and AC are plotted in the diagram. The data for porphyrin amounts in urine of the two samples of each individual in the HCCP and AC group differed considerably. This also becomes evident from the correlationcoëfficients of the two samples, which amounted to r (AC) = 0.27 and r (HCCP) = 0.26 respectively.

DISCUSSION AND CONCLUSIONS

A. AC might be a porphirinogenic compound, considering the fluorescence observed in the animals exposed to 200 mg/kg AC. In the positive control group, with quail exposed to HCB, the liver fluorescence was not detected as clearly as expected. Probably by the administration of olive oil in the dosage used, the oil in the liver did fade the fluorescence of porphyrins. The quail experiments did not give any indication that HCCP, ECH or endrin might be porphyrinogenic.

B. In the exposed groups no increase in the urinary porphyrin level was found, when compared with the control group. The correlation coëfficient of the two samples of the same group was very low due to the fluctuation in porphyrin excretion of each individual from day to day.

All porphyrin patterns were normal. A concentration of 100 μg/l is the upper limit of the range (30-100), in which porphyrin concentrations in urine are considered to be normal[6]. Porphyrin concentrations in urine higher than 100 μg/l were found in the exposed groups as well as in the controls. However, the porphyrin patterns of all tested persons were normal.

Apparently due to the negative results of the experimental animal tests and the normal urinary porphyrin patterns of the workers urinary porphyrins are not a suitable parameter for longterm exposure to HCCP, ECH and endrin. In spite of the possible porphyrinogenicity of AC, determined in the quail, the porphyrin amount in urines of the AC workers was normal. This might be explained by the relatively low concentrations to which these workers were exposed.

ACKNOWLEDGEMENTS

The authors are very grateful for the technical assistance provided by the Technological Department, the drawnings by C. Rijpma and M. Schimmel and the photography by A. van Baaren of the Biotechnion of the Agricultural University at Wageningen. They also wish to thank the workers and the Department of Occupational Health Shell Nederland Chemie B.V. who cooperated in this study. We are indebted to Drs. C. Kan of the Spelderholt Institute for Poultry Research who provided the housing of the quail in cages specially adapted for use with volatile compounds. The Department of Veterinary Pharmacology and Toxicology kindly provided the Japanese quail. The laboratory assistance of Miss E.G.M. Harmsen is greatly appreciated. Typing of the manuscript was done by Miss G. van Steenbergen and Mrs. L. Muller Kobold - de Lagh.

REFERENCES

1. Vliet, P.W. van and Strik, J.J.T.W.A. (1977) Porfyrines in de urine als parameter voor de belasting met epichloorhydrine. Internal Report, Department of Toxicology, Wageningen.
2. Wiel-Wetzels, W.A.M. van der and Strik, J.J.T.W.A. (1978) Porfyrines in urine na blootstelling aan endrin. Internal Report, Department of Toxicology, Wageningen.
3. Nagelsmit, A. and Strik, J.J.T.W.A. (1978) Porfyrines als parameter voor de belasting met allylchloride. Internal Report, Department of Toxicology, Wageningen.
4. Wielard, M.J. and Strik, J.J.T.W.A. (1978) Porfyrines als parameter voor de belasting met hexachlorocyclopentadiene. Internal Report, Department of Toxicology, Wageningen.
5. Strik, J.J.T.W.A. (1973) Experimentele leverporfyrie bij vogels. Thesis, Utrecht.
6. Lange, C.E., Block, H., Veltman, G. and Doss, M. (1976) Urinary porphyrins among PVC-workers. Porphyrin in human diseases, Proc. of the first International Porphyrin Meeting, 352, Karger, Basel.
7. Chalmers, J.N.M., Gillam, A.E., Kench, J.E. (1940) The Lancet, 28, 806.
8. Ockner, R.K. and Schmid, R. (1961) Nature, 189, 499.
9. Industrial Hygiene and Toxicology. Vol. Toxicology, Patty, F.A. (ed.) (1967).
10. Jager, K.W. (1970) Aldrin, dieldrin, endrin, telodrin, an epidemiological and toxicological study of longterm occupational exposure. Thesis, Elsevier, Amsterdam.
11. Strik, J.J.T.W.A. (1977) Porphyrinogenic action of polyhalogenated aromatic compounds with special reference to porphyria and environmental impact. Proc. Int. Symp. Biochem. Diagnosis and Therapy of Porphyrias and Lead Intoxication, M. Doss (ed.), Marburg, 151.
12. Doss, M. and Schmidt, A. (1971) Z. Klin. Chem. u. Klin. Biochem., 9, 415.
13. Doss, M. (1974) Porphyrins and porphyrinprecursors. In: M. Curtius and M. Roth (eds.). Clinical Biochemistry, Principles and Methods, Vol. II, 1325 and 1341.

Chemical Porphyria in Man, J.J.T.W.A. Strik and J.H. Koeman eds.

PORPHYRINS IN URINE OF YUSHO PATIENTS

J.J.T.W.A. STRIK[a], H. KIP[a], T. YOSHIMURA[b], Y. MASUDA[c] and E.G.M. HARMSEN[a]

[a]Department of Toxicology, Agricultural University, Wageningen, The Netherlands

[b]Department of Public Health, Kyushu University, Fukuoka, Japan

[c]Daiichi College of Pharmaceutical Sciences, Fukuoka, Japan.

SUMMARY

An annual survey of 16 Yusho patients exposed to rice-oil contaminated with polychlorinated biphenyls (Kanechlor) was extended with urinary total porphyrin and porphyrin pattern analysis. No significant difference was found in the liver function tests and the other parameters studied between the controls and the exposed people.

INTRODUCTION

In Japan in 1968 a lot of PCBs were accidentally mixed with rice-oil. The most important component was "Kanechlor".

About 15.000 people were exposed. After a year the first symptoms of chloracne were identified. The aim of this project was to detect a possible correlation between urinary porphyrin excretion and blood parameters normally used for liver function tests.

The blood parameters were analysed by Dr. T. Yoshimura M.D., Department of Public Health, Kyushu University, Japan.

MATERIALS AND METHODS

From 31 urine samples, sent by dr. T. Yoshimura, 16 were from people exposed to polychlorinated biphenyls in 1968. The other 15 were controls from Japanese not exposed to PCB.

The following analyses were made:

(a) total porphyrin content,

(b) porphyrin pattern,

(c) creatinine.

Methods are described in procedure for porphyrin assessment (this volume).

The samples were stored at -30 °C until determination.

RESULTS AND DISCUSSION

Only 3 out of 16 samples of Yusho's have a total porphyrin level over 100 μg/l (table 1 and 2). According to Doss[1] 100 μg/l is taken as normal value.

Coproporphyrin dominated in all porphyrin patterns of the contaminated persons. Uroporphyrin was also found but less than coproporphyrin (table 3 and 4).

Pathologic porphyrins were present in trace amounts. Only in two cases a chronic hepatic porphyria Type A could be diagnosed, one in the exposed and one in the control group.

In general blood parameters did not show increased values. Only one person (table 4), Number 29 of the control group, showed an increased GPT, γ-GT, amylase and cholesterol.

CONCLUSION

No differences in urinary porphyrin levels, porphyrin patterns and different blood parameters were found between Yusho patients and Japanese controls. Our data are in accordance with Okuda et al.[4] and Nonaka et al.[this volume].

ACKNOWLEDGEMENTS

The authors are very grateful for the technical assistance provided by the Technological Department, the drawnings by C. Rijpma and M. Schimmel and the photography by A. van Baaren of the Biotechnion of the Agricultural University at Wageningen. Typing of the manuscript by Miss. G. van Steenbergen and Mrs. L. Muller Kobold - de Lagh is greatly appreciated.

Prof. Hiroshi Ibayashi, chief of Yusho Study Group at Kyushu University, is acknowledged for his warmest support to this study.

REFERENCES

1. Doss, M. (1971) Klin. Wschr. 49, 939.
2. Doss, M. (1974) B. Porphyrins and Porphyrin Precursors, in H.Cl. Curtius and M. Roth (eds.), de Gruyter, Berlin, Vol. 2.
3. Martinez, C.A. and Mills, G.C. (1971) Clin. Chem., 17, 199.
4. Okuda, T., Nakajima, H., Yatsuki, K., Amano, M. and Umeda, G. (1978) Brit. J. Industr. Med., 35, 61.
5. Strik, J.J.T.W.A. (1973a) Experimentele leverporfyrie bij vogels, Thesis.
6. Strik, J.J.T.W.A. (1977) Porphyrinogenic action of polyhalogenated aromatic compounds, with special reference to porphyria and environmental impact, in M. Doss, Proc. International Symp. Clin. Biochem., Diagnosis and therapy of porphyrias and lead intoxication, Marburg, p. 151.

TABLE 1

URINARY PORPHYRIN LEVELS IN JAPANESE "YUSHO" PATIENTS

Urine number	Sex	Age	Recovery %	Porphyrin quantity (µg/l)	Creatinine (10^{-3} mmol/l)	Porphyrin (µg/mmol creatinine)
1.	M	35	62	29	66	4.4
2.	F	23	63	82	118	6.7
3.	F	29	65	147	120	12.2
5.	F	29	62	61	60	10.2
7.	M	31	58	97	182	5.3
8.	M	23	66	85	218	3.9
13.	F	24	53	117	194	6.0
14.	M	22	53	76	156	4.8
16.	F	35	58	67	86	7.8
17.	F	21	66	87	226	3.8
19.	F	34	62	57	38	15.1
20.	M	30	63	51	104	4.9
27.	F	25	47	132	124	10.6
28.	M	31	66	36	122	2.9
29.	M	29	65	85	104	8.1
31.	M	38	53	45	45	10.1
average			61	78	123	7.3
standard deviation				29		3.4

TABLE 2

URINARY PORPHYRIN LEVELS IN JAPANESE CONTROLS

Urine number	Sex	Age	Recovery %	Porphyrin quantity (µg/l)	Creatinine (10^{-3} mmol/l)	Porphyrin (µg/mmol creatinine)
4.	M	31	55	85	220	3.9
6.	M	32	66	78	144	5.4
9.	F	27	47	46	86	5.4
10.	F	23	66	57	158	3.6
11.	M	35	65	37	140	2.6
12.	M	38	56	24	98	2.5
15.	F	36	66	41	74	5.6
18.	M	31	64	30	191	1.6
21.	F	27	65	96	82	11.8
22.	M	31	55	125	262	4.8
23.	F	25	62	18	16	11.5
24.	M	35	66	120	211	5.6
25.	F	26	58	87	156	5.6
26.	M	29	66	57	190	3.0
30.	F	37	56	55	108	5.1
Average			61	64	142	5.2
Standard deviation				32		2.3

TABLE 3

QUALITATIVE PORPHYRIN PATTERN IN URINES OF YUSHO PATIENTS AND CONTROLS

Code number	Total porphyrin quantity (μg/l)	Porphyrin pattern (number of methyl esters)*					Type of chronic hepatic porphyria
		4	5	6	7	8	
A. Cases							
1.	29	2				1	N
2.	82	8				1	N
3.	147	28	2		5	1	A
5.	61	2				1	N
7.	97	1	tr.	tr.	tr.	1	N
8.	85	6	tr.		tr.	1	N
13.	117	2				1	N
14.	76	6	tr.	tr.	tr.	1	N
16.	67	5	0.2			1	N
17.	87	2	tr.			1	N
19.	57	7				1	N
27.	132	10	tr.		tr.	1	N
29.	85	7	tr.	tr.	tr.	1	N
31.	45	8	tr.		tr.	1	N
B. Controls							
4.	85	2				1	N
6.	78	35				1	N
21.	96	1				1	N
22.	125	3	0.5		0.5	1	N
23.	18	tr.				-	N
24.	120	3	0.5		1	1	A
25.	87	2	0.5		0.5	1	N

Abrevations:

4 = coproporphyrin 5 = pentacarboxylic porphyrin 6 = hexacarboxylic porphyrin
7 = heptacarboxylic porphyrin 8 = uroporphyrin

* relative visual estimation tr. = trace

TABLE 4

PREVALENCE OF SUBSTANCES IN LIVER FUNCTION TESTS IN YUSHO PATIENTS AND CONTROLS

Test	Control $\bar{x}$	Case $\bar{x}$	Normal range values	
Total Bilirubin	0.6	0.4	< 1.1	mg/dl
Glutamic Oxaloacetic Transaminase	19	19	7-38	units
Glutamic Pyruvic Transaminase	16	17	1-30	units
Alkaline Phosphatase	5.0	4.7	2.8-10.5	units
Leucine Amino Peptidase	156	173	100-210	units
γ-Glutamyl Transpeptidase	16	22	< 50	g/ml
Lactic Dehydrogenase	284	278	190-400	units

Chemical Porphyria in Man, J.J.T.W.A. Strik and J.H. Koeman eds.

ANALYSIS OF URINARY PORPHYRINS IN POLYCHLORINATED BIPHENYL POISONING (YUSHO) PATIENTS

S. NONAKA, T. SHIMOYOMA, T. HONDA and H. YOSHIDA

Department of Dermatology, Nagasaki University, School of Medicine, Nagasaki, Japan. (Director: Prof. H. Yoshida)

SUMMARY

Levels of coproporphyrin and uroporphyrin were determined in the urine of 71 patients with polychlorinated biphenyl (PCB) poisoning and 180 persons who were living in the same area.

The mean level of coproporphyrin of the patient group was 29.4 µg/l and that of control group was 30.6 µg/l. Urinary uroporphyrin was detected in only one patient, who was clinically suspected of porphyria cutanea tarda. However, the significance of PCB poisoning as a pathogenic factor for this case of porphyria cutanea tarda remains obscure.

INTRODUCTION

In 1968, a sudden outbreak of polychlorinated biphenyl (PCB) poisoning (YUSHO) occurred in a Western area of Japan. At that time, approximately 550 patients were found in Nagasaki Prefecture. Some authors have pointed out that abnormal porphyrin metabolism can be caused by ingestion of polyhalogenated aromatic compounds, and it has also been shown in animal studies, that porphyrin levels in the urine are elevated by oral administration of these compounds. These findings suggested that there might be some possibility of high occurrence of porphyrinuria among the Yusho patients and people in the same area. This communication has been based on a record of periodical medical examination in Yusho, in an attempt to find any abnormal porphyrin metabolism.

MATERIAL AND METHODS

Serum and urine samples were collected at periodical medical examinations of 71 patients and 180 persons living in the same area; they were stored at -20 $^{\circ}$C.

Coproporphyrin levels in the urine were determined by the modified method of Sano et al.[8] by ethyl acetateacetic acid (3:1 v/v extraction). Urinary uroporphyrin levels were determined by the method of Dressel[1]. Red blood cell count, chemical determination of serum GTP, GOT and LDH were done at the laboratory of Goto Central Hospital. The serum PCB was also determined by the alkaline hydrolization method.

RESULTS

The results obtained at the medical examinations are shown in tables 1 to 3. The urinary coproporphyrin level of the patients was 26.4 $\pm$ 20.6 μg/l, and of the controls was 32.3 $\pm$ 28.3 μg/l. There was no difference between groups in urinary coproporphyrin. Urinary uroporphyrin was not detected, however, there were some possibilities that the urinary samples were too small to detect the uroporphyrin. There was no correlation between urinary coproporphyrin contents, PCB concentration of serum and liver function.

TABLE 1

THE DISTRIBUTION OF AGE IN PATIENTS WITH POISONING (YUSHO)

Year	Male	Female	Total
0 - 4	0	0	0
5 - 9	5	1	6
10 - 14	5	3	8
15 - 19	4	2	6
20 - 29	2	6	8
30 - 39	3	5	8
40 - 49	9	6	15
50 - 59	4	6	10
60 -	8	2	10
Total	40	31	71

TABLE 2

THE DISTRIBUTION OF AGE IN CONTROLS

Year	Male	Female	Total
0 - 4	0	0	0
5 - 9	5	7	12
10 - 14	4	15	19
15 - 19	7	3	10
20 - 29	4	4	8
30 - 39	1	12	13
40 - 49	10	32	42
50 - 59	9	28	37
60 -	6	33	39
Total	46	134	180

TABLE 3

URINARY PORPHYRINS AND SOME SERUM CONSTITUENTS FROM YUSHO PATIENTS AND CONTROLS

	Patients with PCB poisoning (Yusho)			Controls	
	Mean	S.D.	Case	Mean	S.D.
Cases	70		1	180	
Age	37.0	20.9	49	41.5	20.5
Urinary Coproporphyrin (μg/l)	26.4	20.6	316.8	32.3	28.3
Uroporphyrin			2514.8		
Serum GOT (unit)	31.3	16.6	19	30.2	31.9
Serum GPT (unit)	20.3	15.7	18	23.1	24.5
RBC ($\times 10^4$)	413.3	40.5	421	424.7	38.1
PCB (ppb)	6.0	4.9		3.3	2.3

The following are the details of the patient who showed elevated urinary porphyrin levels.

Case: a 49-year-old man was seen on November, 21, 1975. There was a long history of alcohol consumption. He first noticed hyperpigmentation in exposed areas seven years ago, and has since occasionally developed small vesicles on the back of the hands. He himself did not notice any photosensitivity. A cholecystectomy was performed for chlolithiasis three years ago. At our first medical examination, there was hyperpigmentation, and there were also several rice-sized vesicles and erosions with crusts on exposed areas. Several scars with brown pigmentation were also seen on his face and the back of his hands. The findings of liver function tests were as follows: total bilirubin 0.9 μg/dl, Thymol's test 4.5 units, Kunkel test 11.6 units, total protein 8.2 g/dl, SGOT 43 RF units, SGPT 29 RF units, and LDH 290 units. The urinary coproporphyrin level was 316.8 μg/l, uroporphyrin level was 2514.8 μg/l. Histological examination of skin and liver could not be performed. These findings were suggestive of porphyria cutanea tarda.

DISCUSSION

Recently, some authors reported that the polyhalogenated aromatic compounds, hexachlorobenzene (HCB)[3,4], polychlorinated biphenyl (PCB) and hexabromobiphenyl (HBB) can evoke hepatic porphyria in experimental animals[9]. Also, hexachlorobenzene-induced porphyria has been reported in Turkey since 1957[5]. Peters[7] reported that more than 3000 cases with a mortality rate of 10 per cent occurred in this area. These hexachlorobenzene-induced porphyrias seem to coincide with porphyria cutanea tarda showing bullae, crusts, scars, atrophy of skin, hyperpigmentation, slcerodermatous changes and portwine-colored urine. Abdominal pain is rarely reported. Experimental porphyria due to hexachlorobenzene has been confirmed in an animal study[9].

Vos et al.[10] reported that PCB was a porphyrinogenic substance. Goldstein et al.[2] reported that Aroclor 1254, which consists of a mixture of PCB's, produced an experimental hepatic porphyria in rats, and that this PCB-induced porphyria is characterized by delayed appearance, increased excretion of urinary uroporphyrin, accumulation of 8- and 7-carboxylicporphyrin in the liver and increased drug-metabolizing capacity of the liver. They reported that the delayed onset and excretion of large amounts of urinary uroporphyrin are similar to those seen in hexachlorobenzene-induced porphyria. Miura et al.[6] reported that fasting abruptly elevated the urinary coproporphyrin and uroporphyrin levels in PCB administered rabbits though urinary coproporphyrin levels in the same rabbits were slightly elevated when the diet was continued.

The study reported in this communication deals with the analysis of urinary coproporphyrin in seventy-one patients with PCB poisoning. All patients were found to have a normal porphyrin metabolism, except one. Liver functions, anemia and PCB concentration of serum did not correlate with urinary coproporphyrin levels. However, this study was done seven years after occurrence of PCB poisoning, therefore it was not certain whether the patients with PCB poisoning had abnormalities in porphyrin metabolism at the beginning of PCB poisoning. It was interesting that a case of porphyria cutanea tarda was suspected among the patients with PCB poisoning. He had cutaneous symptoms, abnormalities of urinary porphyrins, but had history of alcohol consumption, therefore the relationship between abnormal porphyrin metabolism and PCB poisoning in this patient was not clear. It is worth describing this case because there has, as yet, been no report of porphyria cutanea tarda being suspected in patients with PCB poisoning.

REFERENCES

1. Dressel, E.I.B., Rimington, C. and Tooth, B.E. (1956) Scand. J. Clin. Lab. Invest. 8, 73.
2. Goldstein, J.A., Hickman, P. and Jue, D.L. (1974) Toxicol. Appl. Pharmacol. 27, 437.
3. Lui, H., Sampson, R. and Sweeney, G.D. (1975) Hexachlorobenzene porphyria. In: Porphyrins in human diseases, M. Doss (ed.), Karger, Basel, p. 405.
4. De Matteis, F. (1967) Pharmacol. Rev. 19, 523.
5. De Matteis, F. (1968) Sem. Hematol. 5, 409.
6. Miura, A. et al. (1973) Jap. J. Hyg. 28, 83. (Japanese)
7. Peters, H.A. (1976) Fed. Proc. 35, 2400.
8. Sano, S. and Granick, S. (1961) J. Biol. Chem. 236, 1173.
9. Strik, J.J.T.W.A. (1973) Enzyme, 16, 224.
10. Vos, J.G., Strik, J.J.T.W.A., Holsteijn, C.W.M. and Pennings, J.M. (1971) Toxicol. App. Pharmacol., 20, 232.

Chemical Porphyria in Man, J.J.T.W.A. Strik and J.H. Koeman eds.

COPROPORPHYRINURIA AND CHRONIC HEPATIC PORPHYRIA TYPE A FOUND IN PEOPLE FROM SEVESO (ITALY) EXPOSED TO 2,3,7,8-TETRACHLORODIBENZO-p-DIOXIN (TCDD)

A.H.J. CENTEN[a], J.J.T.W.A. STRIK[a] AND A. COLOMBI, M.D.[b]

[a]Department of Toxicology, Agricultural University, Wageningen, The Netherlands

[b]Volunteer Researcher, belonging to the Scientificial Popular Comitee, Milano, Italy

SUMMARY

The urinary porphyrin pattern appears to be a more sensitive indicator for chronic exposure to TCDD than total porphyrin excretion. The latter only becomes meaningful at levels higher than 200 µg/l, whereas an abnormal pattern may coincide with levels below this value.

PURPOSE

The aim of the project was to determine first if urinary porphyrins can be used as a parameter for human exposure to TCDD and secondly to see if the data obtained from porphyrin measurements can be related to the degree of environmental exposure of the persons involved.

INTRODUCTION

Chronic exposure to polyhalogenated aromatic compounds may cause hepatic porphyria[1]. In chronic hepatic porphyria (CHP) the pattern of excretion of porphyrins in the urine and their accumulation in the liver clearly suggest the presence of one or more enzymatic defects, which are most likely to be found in the decarboxylation of uro- and heptacarboxylic porphyrinogens[2]. Doss defines porphyria cutanea tarda (PCT) as clinically overt CHP, whereas the clinically occult forms are divided into types A, B and C according to the amount and distribution of porphyrins in the liver and urine. Type C is clinically latent PCT, and CHP Type A, the mildest form, marks the beginning of progression. Type B is characterized by inversion of the normal ratio of coproporphyrin (copro) to uroporphyrin (uro) in the urine, with uro as the dominant porphyrin[2].

Two enzymatic defects are responsible for the development of CHP: increased availability of delta-aminolevulinic acid due to enhanced activity of delta-aminolevulinic acid synthase and limiting of the decarboxylation of heptacarboxylic porphyrinogen. Differences in the regulative, quantitative balance between these two mechanisms determine the various patterns of porphyrin excretion in the three forms of CHP[3]. See table 1.

TABLE 1

DISTRIBUTION OF PORPHYRINS AND TOTAL PORPHYRINS IN URINE IN HEPATIC PORPHYRIAS

Chronic hepatic porphyria (CHP)	c/u-ratio	u (%)	7 (%)	Components	Total Porphyrins in urine (μg/l)
Normal	2-6	15-50	< 3	c>u>>>(traces of 7,6,5)	0-200*
Coproporphyrinuria	> 6	15-50	3	c>u>5>7>6	100-200
Type A_1	> 6	15-50	5-15	c>u>5>7>6	200-600
Type A_2	> 6	15-50	5-15	c>u>7>5>6	200-600
Type B	< 1	> 50	15-20	u>c>7>5>6	200-600
Type C	<< 1	> 50	20-30	u>7>c>5>6	400-1400
Type D (Porphyria Cutanea Tarda) (PCT)	<<< 1	45-80	25-35	u>>7>>>c(5,6) >5(c,6) >6(c,5)	600-1500

* This volume

Abbreviations: u and c = uro- and coproporphyrin,
7,6 and 5 = hepta-, hexa-, and pentacarboxylic porphyrins

The Seveso case

In the village of Seveso, in northern Italy, an industrial explosion contaminated a large area with TCDD. Under certain processing conditions TCDD, 2,3,7,8-tetrachlorodibenzo-p-dioxin, may be formed as an impurity during the synthesis of the herbicide 2,4,5-trichlorophenoxyacetic acid. An estimated 1 to 5 kg of TCDD were released into the environment on July 10, 1976 from the Icmesa chemical plant. Later that year the polluted area was divided into three zones: zone 'A', TCDD concentration higher than 50 $\mu g/m^2$, zone 'B', TCDD concentration between 5 and 50 $\mu g/m^2$ and zone 'R', TCDD concentration lower than 5 $\mu g/m^2$. See also the figures 1, 2 and 3 of the next paper.

It is well known that TCDD may cause porphyria (and chloracne) in workers involved in the manufacture of 2,4,5-T[4,5,6]. TCDD may also cause a number of other toxic effects in man and laboratory animals.

MATERIAL AND METHODS

The 115 urine samples were collected at the end of March 1978. The urine was stored frozen in 50 ml polyethylene bottles and transported to the Netherlands in dry ice. The samples were taken from the following groups of persons: Icmesa workers, people living in the quarters Fanfani and Barruccana as well as from people evacuated from zone 'A' now living in zone 'B' and the quarter Casina Savina. Their location is shown in the figures given by Colombi (this volume). The TCDD contamination of these groups is summarized in table 2.

The total porphyrin determination has been carried out according to the method of Doss[9]. The porphyrin pattern has been estimated by analytical thin-layer chromatography of the methylesters of the porphyrins[9].

RESULTS

In the scatter diagram (fig. 1) the data for total porphyrin levels are compiled. 22,6% of the samples have a porphyrin level above 100 µg/l. The average of the 115 samples is 77 µg/l. The averages of the different groups are as follows: Casina Savina 80 µg/l; Barruccana 76 µg/l; Evacuated 85 µg/l; Icmesa's workers 76 µg/l and Fanfani Houses 74 µg/l.

TABLE 2

TCDD CONTAMINATION OF THE GROUPS

Pollution Zone (officially defined)	Territorial definition of the groups studied
zone 'A' most polluted poll. above 50 $\mu g/m^2$	a. People evacuated, some went back to decontaminated houses in October 1977
Zone 'B' poll. between 5-50 $\mu g/m^2$	b. People evacuated from 'A' living now in Zone 'B'
	c. Barruccana
Zone 'R' officially defined 'Respect'; poll. under 5 $\mu g/m^2$ but where animals died and people had symptoms	d. Fanfani Houses e. Casina Savina
Living scattered in Zones 'B' and 'R'	f. Icmesa's workers

The porphyrin distribution of the 96 samples examined is shown in Table 3.

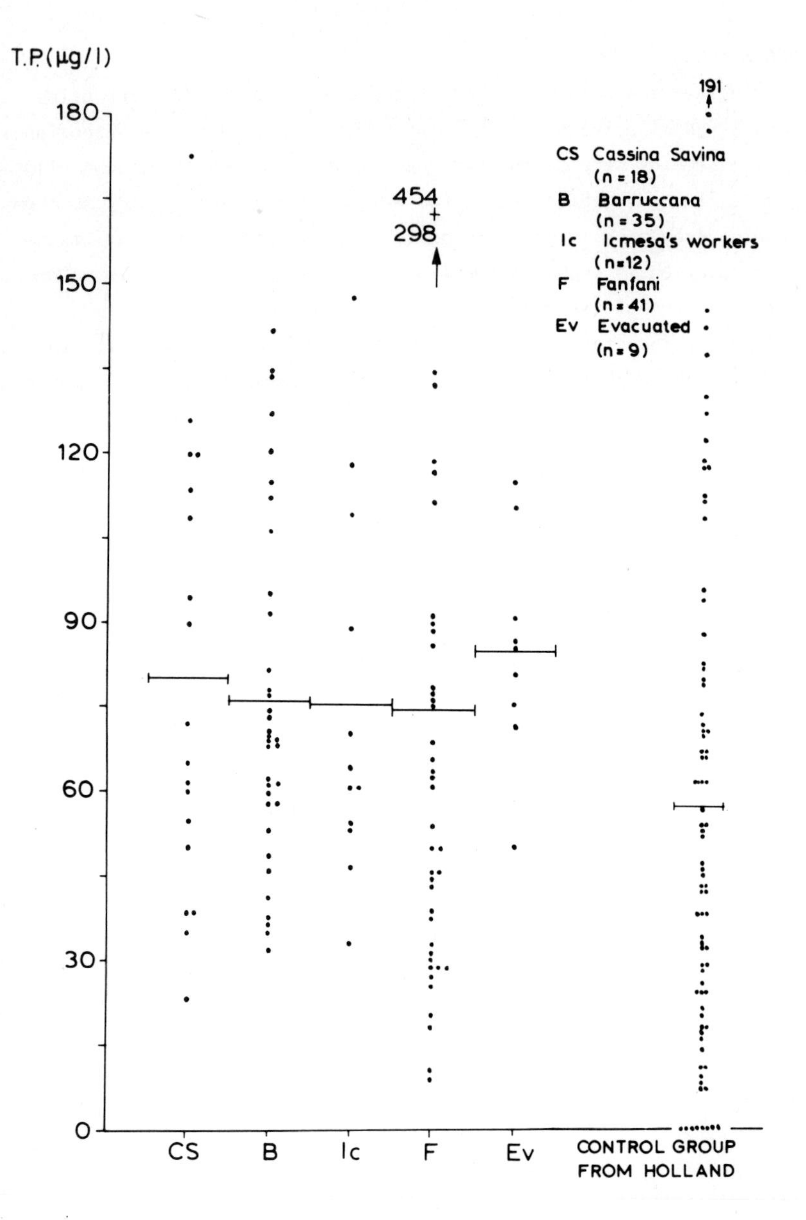

Fig. 1. Total urinary porphyrin values of different social groups from Seveso-area.

TABLE 3

DISTRIBUTION OF PORPHYRINS IN URINE OF TCDD-EXPOSED PEOPLE

Components-distribution	% of samples[b]
c>u>> (traces of 7,6,5)[a]	16
c>u>5≥7>6	36
c>u>7>5>6	48

[a]c = coproporphyrin
u = uroporphyrin
7 = 7-COOH-porphyrin
6 = 6-COOH-porphyrin
5 = 5-COOH-porphyrin

[b]96 randomized samples were run for TLC. Some samples contained high total porphyrin levels, whereas others contained low levels.

The larger part of the people examined belong to the group showing an abnormal porphyrin pattern: 84%.

TABLE 4

PORPHYRIN PATTERN OF THE DIFFERENT SOCIAL GROUPS

Groups	% samples examined with a component-distribution		
	c>u>> (traces of 7,6,5)	c>u>5≥7>6	c>u≥7>5>6
Casina Savina	50	21	29
Barruccana	21	38	41
Evacuated	0	50	50
Icmesa's workers	0	33	67
Fanfani Houses	6	38	56

In this case the porphyrin-pattern shows a difference between the different social groups (table 4).

DISCUSSION

Urinary total porphyrin levels of the groups are not different. In comparison with the Dutch controls there is a slightly higher value. Controls from Wisconsin (U.S.A.) are higher than the exposed people in Seveso (see this volume). Table 3 shows that in 48% of the people examined CHP Type A_2 was found.

Jirasek[5] who analysed 24 hr. urine from factory workers exposed to TCDD found a total porphyrin level of 172-2230 μg/l. The porphyrins-patterns in the present study showed increased values of uro- and 7-COOH-porphyrin. The data from Table 4 indicate that a dose-effect relationship exists as the porphyrin-patterns are in accordance with the degree of environmental pollution (see the map of the polluted area).

CONCLUSION

At the time the observations were made the people exposed to TCDD did not excrete increased amounts of porphyrins, but the porphyrin-pattern was abnormal in many cases. According to Doss[2], people examined suffer from chronic hepatic porphyria type A_2. The effects found in the different groups could be related to the degree of contamination of the parts of the environment involved. It again appears that the urinary porphyrin-pattern is a more sensitive indicator for porphyria than total porphyrin excretion.

ACKNOWLEDGEMENTS

The authors are very grateful for the technical assistance provided by the Technological Department, the drawnings by C. Rijpma and M. Schimmel and the photography by A. van Baaren of the Biotechnion of the Agricultural University at Wageningen. We are indebted to Miss E.G.M. Harmsen for her laboratory assistance.

Dr. Doss reviewed the preliminary internal report and agreed on the findings that alterations in urinary porphyrin in pattern may be used as an indication for a subclinical toxic process.

Typing of the manuscrips by Miss G. van Steenbergen and Mrs. L. Muller Kobold-van Lagh is greatly appreciated.

REFERENCES

1. Strik, J.J.T.W.A. (1973b) Enzyme, 16, 224.
2. Doss, M., Schermuly, E., look, D. and Henning, H. (1976) Enzymatic Defects in Chronic Hepatic Porphyrias in: M. Doss (ed.). Porphyrins in Human Diseases, Karger, Basel, p. 286.
3. Doss, M. (1971) Klin. Wschr., 49, 941.
4. Bleiberg, J., Wallen, M., Brodkin, R. and Applebaum, J.L. (1964) Arch. Derm., 89, 793.
5. Jirásek, L., Kalensky, J., Kubec, K., Pazderova, J. and Lukas, E. (1964) Hautartzt, 27, 328.
6. Poland, A., Smith, D., Metter, C., Possick, P. (1971) Arch. Environm. Health, 22, 316.

7. Poland, A. and Kende, A. (1976) Fed. Proc., 35, 2404.

8. Goldstein, J.A., Hickman, P., Bergman, H. and Vos, J.G. (1973) Res. Commun. Chem. Pathol. Pharmacol., 6, 919.

9. Doss, M. (1974) Porphyrin and Porphyrin precursors, in: M.C. Curtius and M. Roth (eds.), Clinical Biochemistry, Principles and Methods, Vol. II, De Gruyter, Berlin, 1339.

Chemical Porphyria in Man, J.J.T.W.A. Strik and J.H. Koeman eds.

SUBJECTIVE SYMPTOMATOLOGY PREVALENCE AS AN ADDITIONAL CRITERION TO DEFINE RISK-GROUPS EXPOSED TO TCDD IN THE SEVESO AREA, ITALY

A.M. COLOMBI
Scientific and Technical Popular Committee of Seveso, Via Marco Greppi 8, 20137 Milan, Italy

SUMMARY

After a brief history of the aftermaths of the July 10th 1976 TCDD escape in Seveso, an evaluation of the inadequacy of the criteria adopted by official authorities on both contamination definition and epidemiological reserach, is given. A volunteer research has been done on groups of the population both excluded from the official health plan, and known to live in risk-areas according to:

- TCDD levels higher than levels officially accepted
- Animals' murrain occurrence positive for TCDD in organs and/or liver
- Territorial distribution of acute dermatological lesions in children less than 5 years old.

Subjective symptomatology was assessed in the riskgroups so defined in order to stimulate official medical care of the populations concerned. Significant differences were found between symptoms prevalence before and after the TCDD event in each group considered, and between groups presumably exposed to different TCDD levels as inferred by the above epidemiological criteria.

INTRODUCTION

Seveso. Brianza di Seveso is an area of about 9.500 hectars, 20 km north of Milan, Italy; it is subdivided into eleven small towns with a total population of 220.000 (Fig. 1).

TCDD accident. On the 10th of July 1976 the content of the trichlorophenol reactor of the Givaudan-ICMESA factory, located in the territory of Meda near its border with Seveso, was discharged directly into the atmosphere and precipitated in the form of a cloud of droplets. This resulted in the local environment, an urban area of about 250 acres, mainly to the south east of the factory, with a population of several thousand people, being contaminated with 650 g[1] to 2 kg (CSTP) of TCDD.

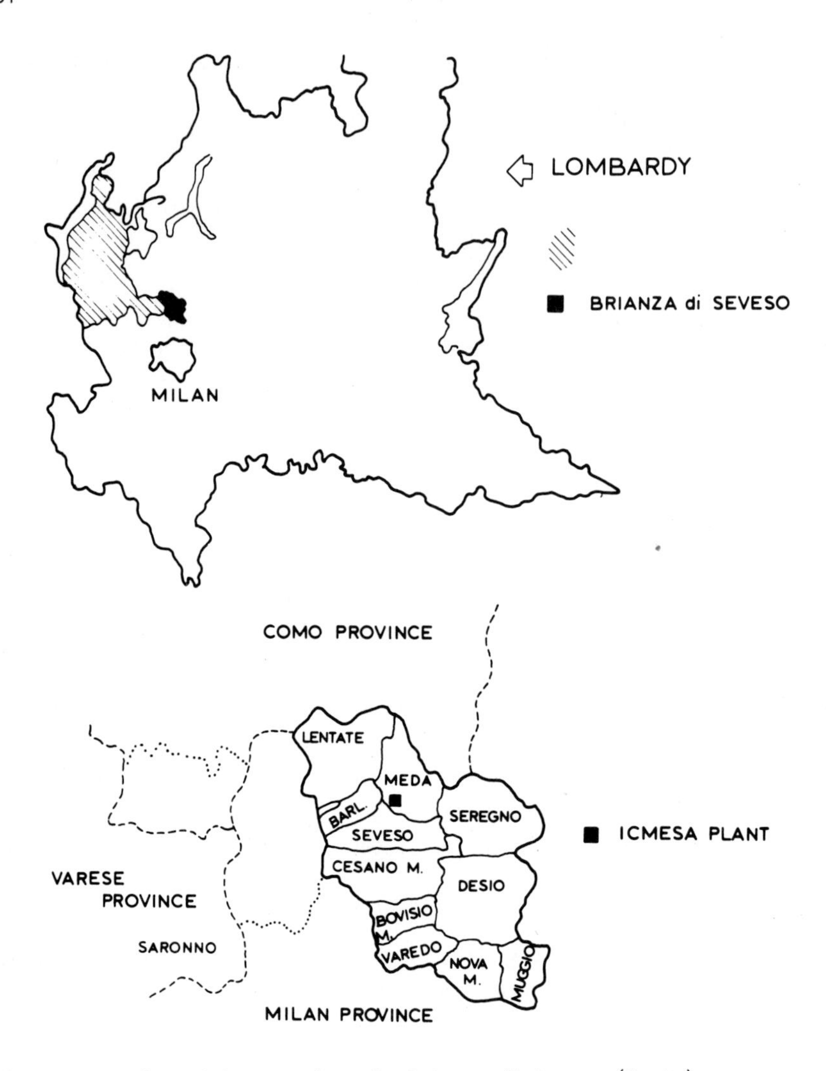

Fig. 1. Location of Icmesa plant in Brianza di Seveso (Italy).

Divisions in zones of the polluted area.

On the basis of the first analytical results produced by Givaudan laboratories the local authorities were formally requested to evacuate immediately in order that the most contaminated area should be placed in quarantine[2]. According to analytical assessments, local authorities divided the contaminated area into 3 zones each with different TCDD levels (Table 1, figure 2)

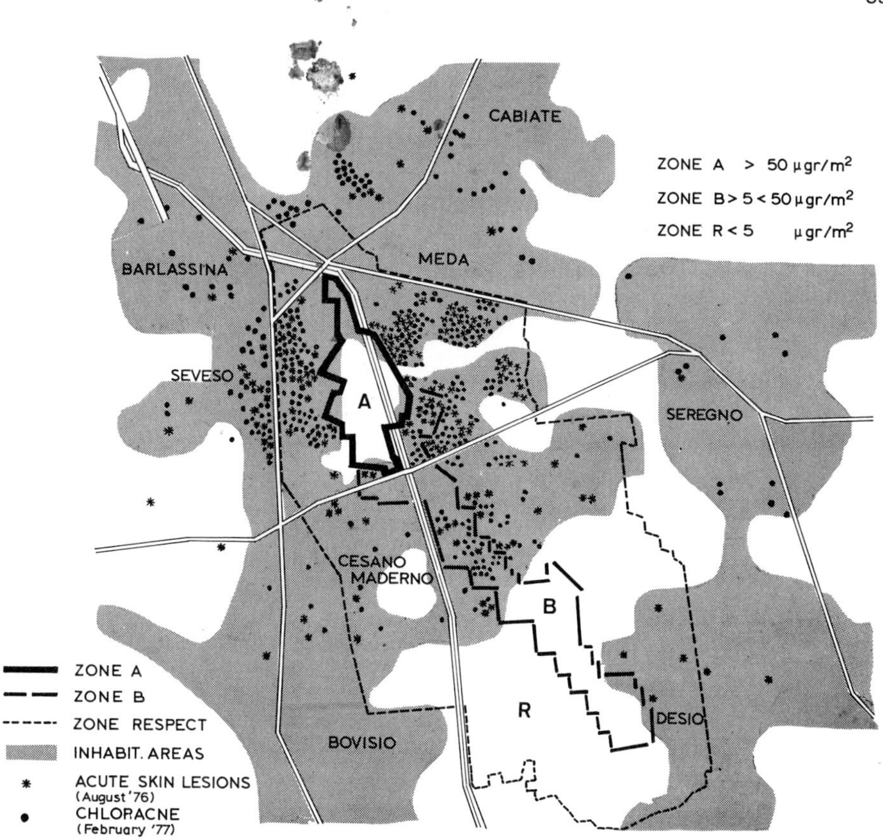

Fig. 2. Definition of TCDD contaminated areas.

- Zone A covering 115 acres with 733 people
- Zone B " 245 " " 5.000 "
- Zone R " 2.500 " " 100.000 "

Finally only the population living in Zone A (733 people) was evacuated. This started on July 25th 1976. For the people remaining in zone B (5,000) recommendations were issued to reduce the likelihood of exposure, while for people living in zone R (100,000) general hygienic rules were suggested.

The decision to evacuate only one part of the population from the contaminated area, leaving about 5,000 people in contact with TCDD, in fact, marked a border between the concentration of TCDD which was considered unsafe and the concentration which was to be conventionally held as not affecting health even in cases of continuous uninterrupted exposure[2].

TABLE 1

SUBDIVISION OF TCDD CONTAMINATED AREA ACCORDING TO CONCENTRATIONS OF TCDD ON THE GROUND

Zone	Pollution values micrograms/sq.meter		Area hectars	Decontamination completed
	Mean	Max.		
A1	580.4	5477	10.7	Zero
A2	521.1	1700	5.1	Zero
A3 (nord)	453.0	2015	3	Zero
A3 (sud)	93.0	441	}	Zero
A4	134.9	902	} 27	Zero
A5	62.8	427	}	Zero
A6	29.9	270	} 62.2	100%
A7	15.5	91.7	}	100%
B	3	43.8	269.4	50%
R	< 5	< 5	1430	just started

A concentration of 5 $\mu g/m^2$ was estimated to be an acceptable value. This was arbitrarily based on the assumption that the resulting average daily intake would be so negligible as to have no adverse effects on health (Table 2).

Consequently, the people living in zone 'R' (100,000) - defined as 'TCDD level under 5 $\mu g/m^2$ - were officially excluded from the health control-programme issued thereafter.

TABLE 2

REGIONAL LAW FOR DECONTAMINATION PROCEDURES AND HEALTH FOLLOW-UP (17TH JANUARY 1977) FOR TCDD POLLUTED AREAS

A zone or place is 'conventionally' considered safe if:

Level of contamination accepted	Type of environment
< 15 g/m^2	agricultural ground
< 5 g/m^2	civil ground
< 0,75 g/m^2	external walls of properties
< 0,01 g/m^2	inside of houses schools etc.

First signs of toxicological damage

Animals. Animals (rabbits, chickens, birds) were dying around the factory within 4 days after the explosion.

Between July '76 and May '77 more than 80,000 animals (80,341) had died, or had been killed[3] after the Authorities' disposition (Table 3).

TABLE 3

NUMBER OF ANIMALS WHICH DIED OR WERE KILLED AFTER AUTHORITIES DISPOSITION IN THE CONTAMINATED AREA

Animals died after TCDD event		Killed after authorities disposition
Zone A	621	1.799
B	335	11.549
R	2.178	63.374
	3.134	76.722

Of all these[3] about 800 cases were submitted to autopsy and in 300 (0.4%) of the latter, TCDD was estimated in liver or other organs.

Only 10 experimental spy-breedings of rabbits - out of the fifteen proposed by the Regional Veterinarian Service and the hundred advanced by public request - were finally settled in different points of Zone 'R'. Partial data only were presented on the kind of animal, its pathology, and the rate of, and territorial distribution of mortality, thus preventing complete knowledge of this first biological indicator of contamination.

Children. During the days immediately after the accident a small number of children directly exposed to the toxic cloud, showed skin rashes and extensive burns due to the action of some caustic material[1,2].

Chloracne started to appear about 4-6 weeks after direct exposure developing in some to acute dermatitis, and showing a tendency towards healing in others. In February 1977 a new outbreak of chloracne appeared in a wider population of children, and in others who had not possibly suffered acute exposure at the time of TCDD escape.

Moreover, indications of TCDD intoxication were found in cows and horses living in contaminated areas so far considered as 'safe'.

According to the first epidemiological data on acute dermatitis in the emergency period, and to the territorial distribution of chloracne in children, the towns of Seveso, Meda, Cesano Maderno, and to a lesser extent, Desio, seemed to be the most heavily contaminated with TCDD; both acute skin lesions and chloracne, however, have been observed in all towns of the zone (Fig. 2) showing that the distinction of zones 'A' - 'B' - 'R', was inadequate to cope with the situation.

Health condition assessment: analysis of official report

After the outbreak of the second epidemic of chloracne, Regional Health Authorities presented a public report on Health Status on May 28th 1977. This remains the only official public report on the subject[3].

- At that time, less than half of the foreseen health plan was fulfilled[4].
- Apart from very crude incidence figures on dermatological disease (Table 4) and a few data on obstretical disease, no meaningful population based morbidity rates have been produced[5].
- It was also clear 'that the background data in the routine national and local statistics either did not exist or were deliberately underreported[2].
- Non conclusive data were given on
  - cytogenetic studies
  - immuno capability in children
  - abortions and malformations.

TABLE 4

SUBDIVISION OF DERMATOLOGICAL LESIONS PRESENTED IN FIGURE 2[6].

Dermatological lesions	number of cases
Suspected chloracne	612
Chloracne 'clinically' accertained	129
Suspicious lesions not 'classifiable'	150
Dermatological lesions not related to chloracne	310
Acute skin lesions observed in the emergency period (July-August 1976)	91

Partial clinical evaluation was presented only for 733 people evacuated from zone 'A' and 360 people living in zone 'B'. 236 people living in zone 'R' were included on special request.

Elaboration of symptomatological interviews was presented only for 602 people of zone 'A' and two samples of people from zone 'R' (236) not taking into account age or sex, and without any correlation with clinical signs or laboratory results, which themselves were presented without any tabulation.

It was impossible to relate exposure and effects and to correlate the data, due to the following facts:

Zone A on 733 people submitted to assessment,
only 624 had a proper 'exposure and symptoms' interview
" 447 had two successive blood samples taken
" 141 had three successive blood samples taken allowing any comparison (see Table 5)
" 345 were submitted to a complete internal medical assessment
" 405 were submitted to neurological assessment
" 310 were submitted to dermatological assessment
" 35 were submitted to oculistic assessment[3].

The conclusion presented was that 16.7% of subjects had 'hepatomegaly'. Conclusions (no tables or correlation between symptoms and environmental status were given) were presented for the following laboratory tests comparing three consecutive samples of 141 subjects.

TABLE 5
INTERPRETATION OF HEMATOLOGICAL TESTS OF THREE CONSECUTIVE SAMPLES OF 141 SUBJECTS OF ZONE 'A'[3]

Test done	Conclusions drawn (exactly) in official report
Transaminases (SGOT-SGPT)	'Increase for males' (age group 15-45)[a]
γ-GT	'Increase for males' (age group 15-45)
	In-/decrease for other age-groups in both sexes[a].
Bilirubin	'No significant variation in all classes for both sexes'[a]
Alkaline phosphatase	"
Creatinin	"
WBCC (no reference to formula)[b]	Not mentioned

[a]Data given without tables, percentages and absolute figures

[b]Total urinary porphyrin assessment was undertaken in the first few months after the accident: nevertheless no data were mentioned.

Similar conclusions can be reached from analysis of the results given concerning:

- 360 subjects from zone 'B' for whom laboratory results, referring to two consecutive blood samples only, were given without tabulation and without mention of any clinical or symptom study
- 236 subjects from zone 'R' (135 subjects living eastward of the factory from which TCDD escaped; 71 - including a 'Fanfani houses' group - living westward of zone 'A') included after public request.

No correlation was given between symptoms and lab. results, apart from a general conclusion on an average 'increase' in 'hepatic size' (35%) and 'decrease' in WBCC (28%)[3].

If these data appeared meager, it appeared, further, that the methods and criteria leading to such conclusions were less even than the results themselves.

In this context it appeared at least inadequate to conclude that 'an overt pathology of liver, gastrointestinal tract, peripheral nervous system, carbohydrate, fat and porphyrin metabolism, all of which are common features in the cases observed in previous accidents, have not been reported for Seveso residents'[1,2].

The obvious difficulties and limitations encountered in research on this topic, especially that undertaken by a volunteer unofficial research group, can, then, be better understood.

Aims

History. The Scientific and Technical Popular Committee was founded on July 18th 1976 on behalf of some ICMESA's workers, local people who were concerned, and a group of technicians and doctors, with the aim of collecting from the population involved all available knowledge concerning scientific and technical aspects of TCDD contamination. Under the circumstances, it appeared necessary to collect direct information, particularly on health status.

A volunteer research was started both to see if any symptomatology changes suggestive of exposure could be traced, particularly in selected groups of the population excluded by the official health plan, and to stimulate authorities to reconsider current criteria adopted relevant to this situation. Because the rate of TCDD contamination in zone 'A'-'B' was already well established attention was focused on zone 'R'.

On the other hand no access was possible to data or research concerning people in zones 'A' and 'B', while the zone 'R' population, excluded from the health plan, was available for participation in the volunteer reserach.

METHODS

Redefinition of zone 'R'

Within zone 'R', it was possible to define different areas, presumably with a higher risk of exposure to TCDD, through the correlation of the following findings:

(1) Presence of TCDD levels eventually found on consecutive checks higher than official safe' levels (5 $\mu g/m^2$)
(2) Presence of high mortality rate in animals in the emergency period and thereafter of animals found positive for TCDD in organs or liver
(3) Presence of acute dermatological lesions in children under 5 years old, related to territorial distribution.

The correlation of the above findings could, in this way, give an indication for epidemiological research. This methodological approach gained favour. It was introduced as a proposal for future epidemiological work[3]. Nevertheless it was afterwards abandoned from the official health control plan.

Risk areas

Accordingly within zone 'R' 7 different situations could be defined which deserved study (Fig. 3) - (Table 6):

Risk groups

From the risk-areas mentioned, people belonging to the following groups were asked to cooperate in an examination:

(1) Fanfani houses (west of zone A_2; see Fig. 3)
(2) Barruccana (east of zone A_7; see Fig. 3)
(3) Cassina Savina (west of zone B; see Fig. 3).

The Polo population, which needed to be studied, because of the coincidence in the area (east of Icmesa plant) of all the above-mentioned criteria, was not available for this investigation, but after public request, it was submitted to control by the Health Authorities.

Within these areas, groups of population were then defined belonging to the same quarter of town, the same block of flats or group of houses, in order to have risk groups homogenous at least as far as general living conditions and territorial exposure risk were concerned. Such 'risk-groups' which voluntarily adhered to the research after public explanation of its meaning, aims and limitations, were finally submitted to symptomatological interview with a proper questionnaire. They were previoulsy instructed neither to minimize, nor to emphasize symptoms felt either in the 2 years before TCDD escape, or thereafter to the moment of interview.

TABLE 6

CRITERIA TO DEFINE RISK-AREAS PRESUMABLY MORE HEAVILY CONTAMINATED WITH TCDD WITHIN ZONE 'R'[3].

Area	Criteria		
	Analytical	Veterinarian	Dermatological
	TCDD levels eventually above 5 $\mu g/m^2$ on consecutive checks	High mortality rate in animals and TCDD found in organs of animals*	High rate of acute dermatological lesions in children less than 5 yrs.old*
(1) eastward of ICMESA: <u>POLO</u>	+ + + (6 samples) > 5 $\mu g/m^2$	+ + (5 breedings positive for TCDD) 2 of which were official spy rabbit breedings*	+ + + 'high' rate (specific rate unknown)
(2) eastward of zone A: <u>BARRUCCANA</u>	- - (TCDD offical value, 1.8 $\mu g/m^2$). No consecutive checks known.	+ + + Very high mortality rate. Nine breedings of rabbits positive for TCDD in liver, 2 of which were official spy-breedings*	+ + + 'high rate' (specific rate unknown)*
(3) eastward of zone B including <u>CASSINA SAVINA</u>	+ + mean values low, but some samples > 200 $\mu g/m^2$ close to highway	+ + + 'elevated' mortality rate; six breedings of rabbits positive for TCDD, two of which were official spy-breedings*	+ 'rare cases' (specific rate unknown)
(4) south of zone B	- - ?	+ + 5 breedings positive for TCDD of which 1 was outside zone R	+ 'rare cases'
(5) westward of zone B	TCDD mean values < 5 $\mu g/m^2$	no data available	+ + + 'elevated' rate (specific rate unknown)

(6) westward of zone A including <u>FANFANI HOUSES</u>	+ 1 sample 8 µg/m²	+ + + cattle breeding positive for TCDD intoxication including embryotoxicity abortigen effects and TCDD in milk	+ + + chloracne distributed in zone R and beyond it in 'consistent' number (specific rate unknown)*
(7) north to <u>ICMESA</u> factory	TCDD values < 5 µg/m²	- - no conclusive documentation of animals pathology: TCDD in liver of one chicken	'rare' cases (special rate unknown)*

* the exact descriptive word used in the official report (28th May 1977[3]) is here reported: no specific incidence figures were given.

<u>Symptom interview</u>

Apart from general data, the questionnaire also recorded the time of residency in the risk-area and the different possibilities of daily exposure (housewives, workers, children playing out-door etc., did the subject eat vegetables, courtyard animals in the emergency period of thereafter, etc.) and finally more than 100 symptoms were explored. Interviews were run by a volunteer health research group composed of 3 medical doctors and a group of medical students, working with the collaboration of properly instructed members of local cultural associations.

Some obvious limitations hindered assessment:

(1) It was decided to collect only symptomatological changes and not to run internal medical checks for the following reasons:

a. the declared aim of the volunteer research was to include interviewed people in an official health control programme and to stimulate adequate preventive measures such as extending decontamination procedures to risk-areas previously defined as 'safe'.

b. the problem was not to induce a private medical solution to a public environmental problem.

c. prevention of any access to laboratory facilities, because of the unofficial character of the research.

(2) The clinical features and symptoms of the intoxication are not sufficiently distinctive to permit clinical recognition beyond epidemiological criteria

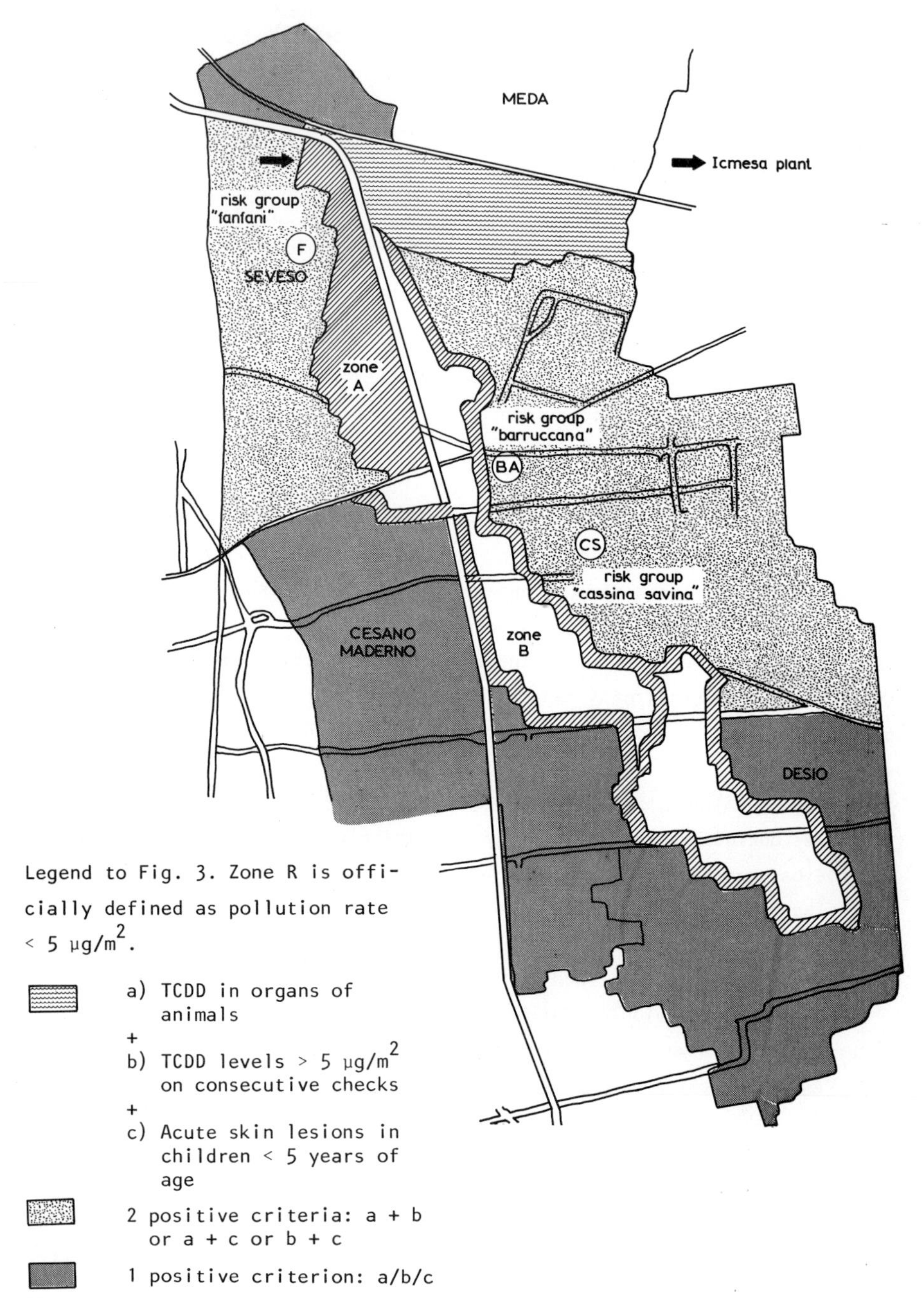

Legend to Fig. 3. Zone R is officially defined as pollution rate $< 5\ \mu g/m^2$.

- a) TCDD in organs of animals
  \+
  b) TCDD levels $> 5\ \mu g/m^2$ on consecutive checks
  \+
  c) Acute skin lesions in children < 5 years of age
- 2 positive criteria: a + b or a + c or b + c
- 1 positive criterion: a/b/c

Fig. 3. Redefinition of Zone R according to epidemiological criteria (see legend)

of exposure. Not only may clinical features abscribed to TCDD concur with other, similar, substances, but many of the symptoms due to these substances might also be induced by numerous other etiological factors, whose importance might be difficult to assess beyond strict epidemiological criteria. Epidemiologically orientated collection and correlation of peripheral data has either been lacking on the official side, or at least not available for public knowledge.

Notwithstanding these obvious limitations, symptomatology was colected on risk-groups to see if any indication of exposure could be found, comparing the prevalence of symptoms not only before and after TCDD event, but also between different groups living in areas related to presumably different rates of exposure as inferred by the above-mentioned criteria.

Collection of data

The first risk-group examined was the population of a block of Council-Houses, named 'Fanfani Houses', in Seveso, about 500 mt. from ICMESA's plant, 10 to 30 mt. from the boundary of the most contaminated zone 'A_2', and about 200 mt. from a cattle breeding which previously showed signs of TCDD intoxication and was positive for TCDD in organs and milk.

Collection of symptomatology of the people living in the 'Fanfani Houses' was done in the period Jan./Febr. 1977 on a total of 71 subjects of both sexes, aged 15-60, out of a possible 191 residents. Symptoms encountered in the 2 years preceeding the TCDD escape, and thereafter to the moment of the interview, were assessed.

Afterwards, the research was implemented on a more standardized basis, focusing on the risk-groups 'Barruccana' (Ba) and 'Cassina Savina' (Cs), which were submitted to interview in the period Dec. 77/Febr. 78. It was assumed that group Ba was exposed to a somewhat higher risk than group Cs according to the epidemiological-environmental findings stated above.

The two available groups could be defined as follows (Table 7):

TABLE 7

Riskgroup 'Ba'	Area featured by
males 33 (age 15-60)	-Impressive animal murrain in the emergency period; 9 breedings of rabbits positive for TCDD in organs
fem. 19 (age 15-29)	
19 (age 30-60)	
children 81 (four age groups)	-High rate of acute dermatological lesions in children under 5 yrs.
total interviewed 152	-TCDD ground mean values officially under 'safe' level (5 $\mu g/m^2$)

Riskgroup 'Cs'	Area featuredby
males 48 (age 15-60)	-Elevated animal murrain but sensibly less than in 'Ba' 6 breedings of rabbits positive for TCDD in organs
fem. 23 (age 15-29)	
55 (age 30-60)	
children 80 (four age groups)	-Dermatological cases in children present but rare
total interviewed 206	-TCDD ground mean values officially under 'safe' level (5 $\mu g/m^2$)

The age distribution of adult males in the two groups was practically the same; the women, on the contrary, had to be divided into two age groups, because the age structure of the samples was different. In order to see if the time of residency, i.e. the time spent daily in the presumably contaminated area, affected the prevalence of symptoms, the population was classified into the following groups:

-persons living all day in the place (mainly housewives)

-persons who worked in the area

-persons who worked elsewhere, but still in zone 'R'.

Symptoms which were frequently reported in literature and whose high incidence was confirmed by official health interviews, were then chosen for further study, leaving apart those, whose relationship with TCDD exposure might be less clear.

The following symptoms were retained, checking for any kind of specific therapy or hospitalization (sp. T. or H.), indicating underlying disease:

-burning and watering of the eyes

-reading test (according to TUNY et al., 1976: loss of distinct reading capability after 5-10 min. in 70% of subj.)

-anorexia

-nausea and/or vomiting
-abdominal pains / colicky pains
-itching
-dermatological lesions / skin spots
-headache
-paresthesias / tingling of extremities
-insomnia
-epistaxis (included as 'spy' indicator of that group of symptoms less clearly abscribable to TCDD exposure).

Thereafter any significant differences between symptoms prevalence in the different risk-groups were studied, expecting that the risk of exposure would be as follows: 'F' > 'Ba' > 'Cs'.

RESULTS

Fanfani.

A general increase in the prevalence of symptoms after TCDD escape was found in comparison to the preceeding period for the 'Fanfani Houses' group. On the other hand, no significant difference was found in the prevalence of symptoms between the present research and the research on the same risk-group reported by Official Health Authorities, after public request (see Table 8).

Also for risk-groups 'Barruccana' and 'Cassina Savina' there was a general increase in symptoms experienced after the TCDD escape as compared to those experienced before it, in both sexes and all age groups.

Symptomatological increase after the TCDD event, appeared higher in risk-group 'Ba', than in group 'Cs'. For instance with males, a significant difference was observed for 4 of the symptoms considered (Fig. 5). In order to discover whether these differences are ascribable to different attitudes of the interviewers or of interview biases, we also looked for differences in prevalence for a series of symptoms bearing no relationship with TCDD. Again with males, the following results were found (Table 9):

TABLE 8

COMPARISON OF SYMPTOMATOLOGICAL PREVALENCE ON THE SAME RISK-GROUP, FANFANI HOUSES

Official local health authorities vs. volunteer research by scientific popular committee, period of observations: jan./mar. 77, total population of 'Fanfani Houses' risk-group: 190[7].

person interviewed[a]	health authorities 56 (group 1)	scientifical pop. com. (volunteer) 71 (group 2)
symptoms observed	% of interviewed	
Conjuctivitis	27.2	22
Nausea	10	8
Astenia	11.5	15
Dyspepsia	11.5	12
Itching	21.5	13
Head ache	6	10
Eyesight disturbances	8	9
Micturition	4	5.5
Objective signs		
- hepatic disease (not related to drugs, alcohol, infective hepatitis)	22.7%	Tendency to lessening of Platelet Count
- skin lesions	1 case chloracne 8 cases acute lesions due to TCDD 6 cases lesions suspected of being due to TCDD	3 pathological cases
- neurological	1 case of peripheral neuropathy	

[a]Different operators tended on the whole to interview different people.

TABLE 9

SYMPTOMATIC CASES VS. EXPECTED IN GROUP 'Ba' FOR 'ASPECIFIC SYMPTOMS'

Symptoms	Chi square	p.value
Cough	0.159	n.s.
Tachycardia	1.565	n.s.
Muscle-cramps	1.572	n.s.
Pollachiuria	0.500	n.s.
Hearing-loss	0.451	n.s.
Cold in extremities	0.355	n.s.
Hair loss	2.384	n.s.
Sore throat	4.372	< 0.05

A significant difference between the two populations was then found only in 1 case.

Thereafter any significant differences between symptoms prevalence in different risk-groups were considered, expecting that the risk of exposure would be as follows: 'F' > 'Ba' > 'Cs'. No reliable comparison could be made between risk-group 'Fanfani' ('F') and the other two groups, because interviews on the former were done about one year before (Feb. '77). Such a comparison could be done instead on groups 'Ba' and 'Cs' interviewed at almost the same time (Dec. 77/Feb. 78).

The prevalence of symptoms observed after TCDD event in group 'Ba' and in group 'Cs' were compared, allowing for the difference in age and sex structure in the two populations. The number of symptomatic cases observed in 'Ba' and those that would have been expected in it, if the age and sex specific prevalence rates were the same as 'Cs' have been assessed, for each age and sex subgroup (Fig. 6) and, in general, for all adults (Fig. 7). The number of expected cases was assessed as:

$$\frac{S_{Cs.i} \times P_{Ba.i}}{P_{cs.i}}$$

where Si = symptomatic cases in the age and sex group considered

Pi = total population of the age and sex group considered

Cs. = risk-group 'Cassina Savina'

Ba. = risk-group 'Barruccana'

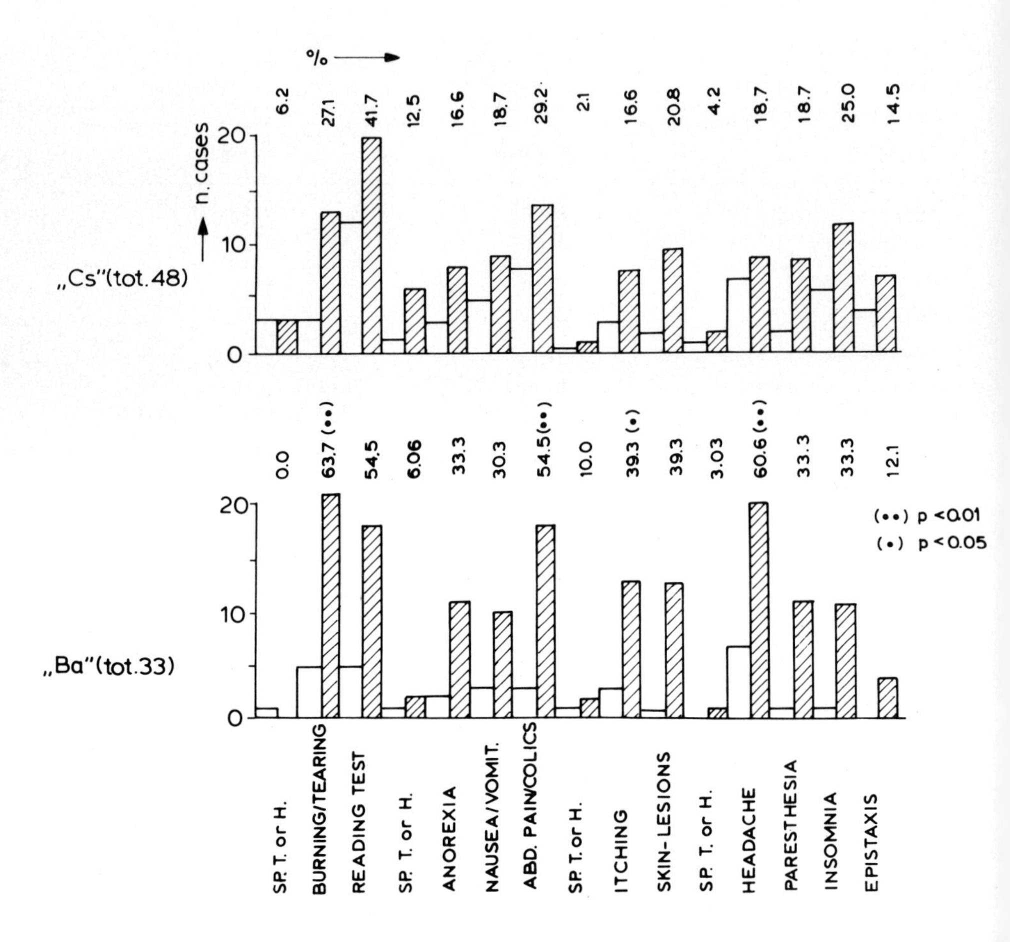

Fig. 5. Orientative distribution of symptomatology before and after TCDD event. Males, ages 15-60. SPT or H = Eventual Specific Pharmacological Treatment or Hospitalization for diseases.

[ ] Number symptomatic cases preexisting TCDD pollution.

[hatched] Number actual symptomatic cases appearing after TCDD event.

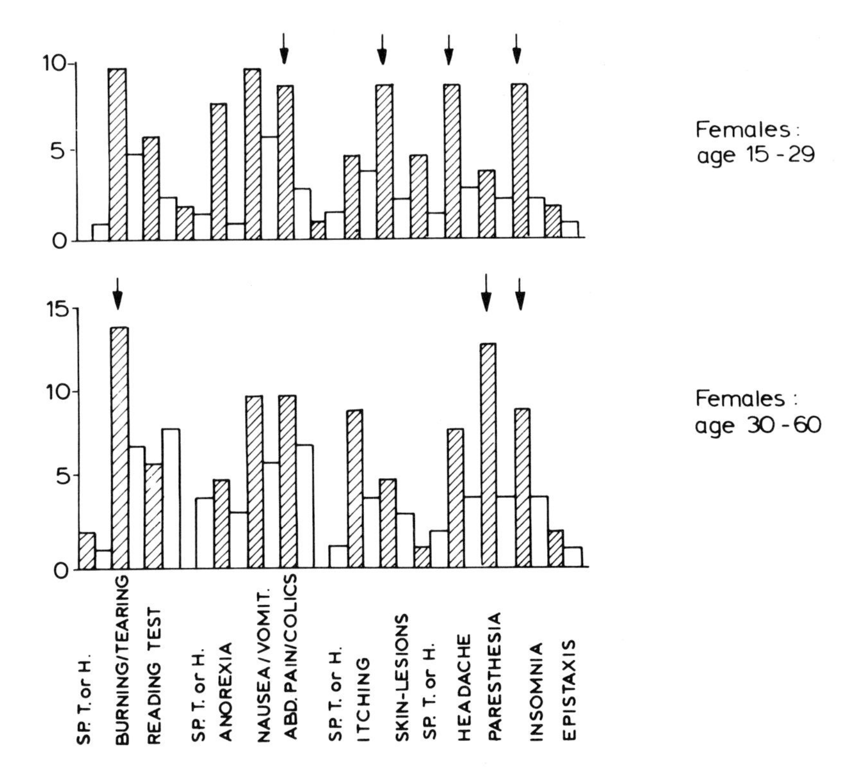

Fig. 6. Symptomatic cases observed versus expected in group Barruccana. Females: age 15-29. Cs. total 23 interviewed and Ba. total 19 interviewed (Jan.-Febr. 1978). Females: age 30-60. Cs. total 55 interviewed and Ba. total 19 interviewed (Jan.-Febr. 1978) (↓) $p < 0.05$

Number of cases observed in risk group Barruccana

Number of cases expected in risk group Barruccana

By applying the summary chi square formula described by Mantel and Haenszel, a significant difference was found for the majority of the symptoms considered. Children were considered in the same way after being divided into the following age groups: 0-3 / 4-6 / 7-10 / 11-15. They showed an increasing difference in symptomatic cases observed vs. expected, with increasing age, which was related to increasing reliability of interview.

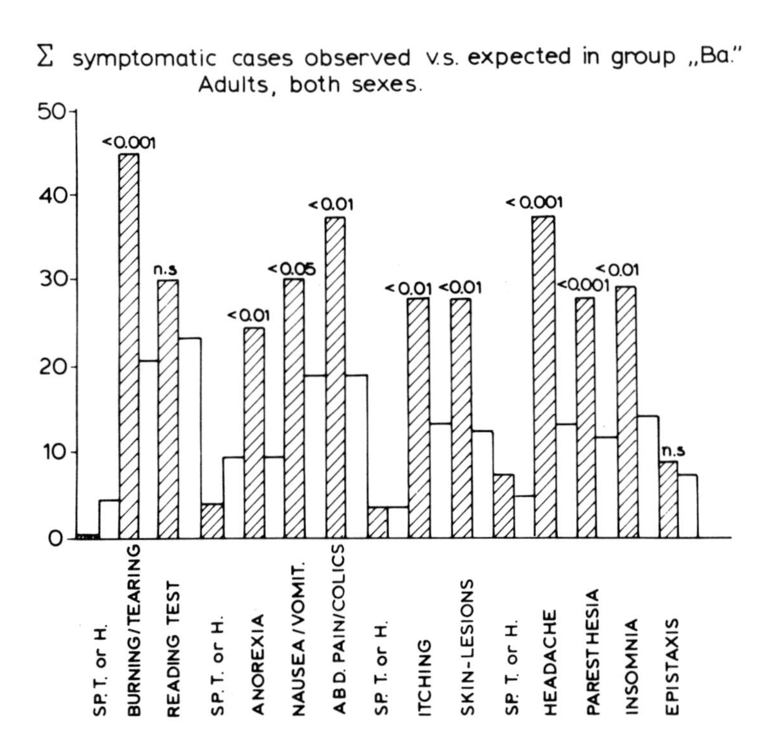

Fig. 7. Total number of symptomatic cases observed versus expected in group Barruccana. Adults, both sexes.

[hatched box] Total number of cases observed in risk group Barruccana.

[open box] Total number of cases expected in risk group Barruccana

n.s = not significant

0.001 }
0.01 } = p value
0.05 }

DISCUSSION

There was no access to centralized data, and, as already explained, no possibility of running laboratory analysis or of undertaking clinical research. However, an attempt was made to show that risk groups of people presumably exposed to TCDD, showed a higher symptomatology after TCDD escape, and that a higher level of this could be expected in higher risk areas defined according to environmental-epidemiological criteria, or, vice versa, that higher prevalence of symptoms could be assumed as an additional criterion of exposure for further epidemiological research in the involved area.

On the whole, the increase in symptomatology felt after TCDD escape in the Fanfani-group was consistent which related topographical and veterinarian data, and also with clinical data available concerning levels of liver disease - not due to alcohol, virus or other toxic substances - and with the decreasing trend in platelet counts.

On the basis of this rough indication, the official Authorities were asked to include these people in the health programme and similarly to revise the health control plan for all contaminated areas irrespective of the so-called 'safe' levels, therefore adopting epidemiological criteria.

As previously stated, this proposal was indirectly accepted by a number of official health officers and was partially included in the general official report[3], indirectly confirming the indicative value of our approach.

Moreover, a direct confirmation of the validity of these data was later given for the same risk-group 'Fanafani houses' by a specific official health report in Nov. 1977[7], in which both the results of this volunteer research and the results of the official body, thereafter issued, were assessed, and shown to have no significant differences (Table 8).

Finally, after the subsequent public request for TCDD re-assessment, it was shown that, whilst previously the Fanfani houses area had been assigned a TCDD value of 0.97 $\mu g/m^2$, consecutive checks revealed outdoor TCDD values of up to 3.56 $\mu g/m^2$.

Nevertheless such an approach has once more been abandoned with the resignation of those who supported it. At this point it was decided to continue the research, on a more standardized basis, on other groups of the population excluded from public care, as was the case for the risk groups 'Barruccana' and 'Casina Savina'. Again, at least indicatively, it has been possible to conclude that in either group, there was a higher symptomatological prevalence in the period after TCDD escape when compared to the preceeding period. Also group 'Ba' showed a significantly higher symptomatological prevalence after TCDD escape, than group 'Cs', supporting the hypothesis that it was exposed to a higher risk.

This observation is consistent with the fact that group 'Ba' is living in an area immediately eastward of zone 'A', where an noticeable animal murrain had taken place and where more chloracne cases happened than in the area related to group 'Cs', immediately eastward of zone 'B', where less dermatological and veterinarian findings could be traced.

If the increase in symptomatology observed in selected risk-groups is ascribable to TCDD exposure and if it correlates positively with environmental-epidemiological dates concerning chloracne distribution, and with veterinarian findings, then an orientative conclusion can be drawn on the necessity that at least all the population living in risk-areas, defined according to the above criteria, should be submitted to more proper and adequate research.

Our experience seems to indicate that only if further research is based on risk-groups so defined, (and if symptoms and signs are correlated with epidemiological and environmental data), will it become possible, for any practical preventive purpose, to see if initial or subclinical signs (including biohumural tests) are significant or not.

The main objections to the collection of subjective symptomatology might be the following:

(1) symptomatology is not feasible for comparison in different groups because data is collected by a heterogeneous group of interviewers. A counterproof that this was not so is given by the comparison in the prevalence of symptoms between group 'Ba' and 'Cs' before TCDD event when risk of exposure is nil for both. In this case no difference could be found between symptoms observed vs. expected in 'Ba' when risk had to be equal to that of 'Cs'.

(2) symptomatology collection is not reliable because differently motivated interviewers will influence interviewed subjects in different ways. A counterproof that this was not so is given by the comparison of symptomatology incidences collected on the same risk-group, 'F' both by the volunteer research group and the Local Health Authorities; no substantial difference was to be found[2].

(3) the increase in the number of symptomatic cases after TCDD escape, is due, not so much to exposure, but more to the expectations induced by the circumstances in the interviewed populations who voluntarily adhered to the research.

Against this last objection it may be said that, with the exception of a few cases, there is no significant difference between the two populations for those symptoms considered to be less clearly related to TCDD exposure. Nevertheless, it would be necessary to determine whether either the symptomatology due to TCDD, in subclinical situations, has a generical wide range, or whether the higher prevalence of symptoms in a higher-risk population has to be ascribed to a higher awareness ot its members for the possible, but not actual, consequences of the contamination. This question will only be solved by correlation with the objective findings of further research.

CONCLUSIONS

Researches on social groups without participation of the same are generally bound to failure as can be seen from the literature and as was admitted in the case of the offical health control programme for Seveso[3]. On the contrary, conscious participation of people seems a quarantee, not a shortcoming, to epidemiological research, especially if it becomes critical, active knowledge, of homogenous risk-groups.

Subjective symptomatology collection is obviously not an exhaustive method of research by itself, but it is useful when correlated with environmental risk evaluation and objective clinical and laboratory signs. In the above context, collection of subjective symptoms in selected risk-groups was a good and inexpensive means, and the only practicable one besides, of adding information for orientation of further research.

Such a methodological approach, which can be used on a homogenous group of workers within a factory, with regard to health, safety and preventative measures, may also be used quite satisfactorily, for similar epidemiological purposes, upon selected groups of individuals outside of the factory.

Collection of subjective symptoms, when related to such groups as a whole, can therefore, when correlated with crude environmental indications of risk, serve as a guide, offering a general orientation for further and more detailed research. This would be especially so in the absence of any other objective official health control programme data, as has been the case with the situation described above.

REFERENCES

1. Reggiani, G. (1977) Toxic effects of TCDD in Man, presented at the Nato Workshop on Ecotoxicology, July-August 1977, Guilford, England.
2. Reggiani, G. (1977) Medical problems raised by TCDD contamination in Seveso - Italy. Paper presented at the 5th International Conference on Occupational Health in the Chemical Industry Medichem, S. Francisco, Cal., Sept. 5-10 1977.

3. Fara, G. (1977) Rapporto Preliminare sullo stato di salute nella zone inquinata da TCDD. Relazione presentata dal Prof. G. Fara presidente della Commissione Medico-Epidemiologica al Convegno di Seveso, 28 Maggio, 1977. In I° Quaderno Documentazioni a cura del Comitato di Coordinamento dei Consorzi Sanitari di Zona Brianza-Seveso 1-2-3.

4. Terracini, B. (1977) Letture: I° Quaderno Documentazioni Rapporto Preliminare sullo stato di salute nella zona inquinata da TCDD in Epidemiologia e Prevenzione Estate 77.

5. Berrino, F. (1978) Preliminary Report on Mortality in the Seveso Area 1975-77 - Milan, Jan. 7, 1978. Presented at the JARC Meeting on the coordination of Epidemiological Studies on the Long-term Hazards of Chlorinated Dibenzodioxins/Dibenzofurans.

6. Bolletino Informazioni N. 4, 3 Nov. 1977 a cura del Comitato di Coordinamento dei C.S.Z. di Brianza-Seveso 1-2-3.

7. Comitato di Coordinamento dei Consorzi Sanitari di Brianza-Seveso 1-2-3. Comitato di Coordinamento Tecnico Documento Protocollo 419/2 del 22 Nov. 1977.

Chemical Porphyria in Man, J.J.T.W.A. Strik and J.H. Koeman eds.

TOXIC PORPHYRINURIA AND CHRONIC HEPATIC PORPHYRIA AFTER VINYL CHLORIDE EXPOSURE IN HUMANS

M. DOSS[a], C.E. LANGE[b] and G. VELTMAN[b]

[a]Department of Clinical Biochemistry, Faculty of Medicine of the Philipp University, Marburg an der Lahn, FRG.

[b]Department of Dermatology, Faculty of Medicine of the Rheinischen Friedrich-Wilhelm University of Bonn, FRG.

SUMMARY

Urinary porphyrin and porphyrin precursors excretion were studied in workers from a PVC (polyvinyl chloride) producing and processing plant. Secondary coproporphyrinuria and partial transition to chronic hepatic porphyria were found consistently in workers with vinyl chloride induced toxic liver damage. The pathobiochemical mechanism for the development of coproporphyrinuria and chronic hepatic porphyria Type A is probably based on a toxic disturbance of hepatic coproporphyrinogen oxidase and uroporphyrinogen decarboxylase by vinyl chloride. In medical supervision of VC (vinyl chloride) exposed workers the determination of urinary porphyrins is a diagnostic and preventive parameter, which indicates toxic injury at an early stage.

INTRODUCTION

Due to the complexity and all-pervasive cultural, biological, and medical significance of the interactions between the environment and porphyrin metabolism, it is not surprising that systematic experimental and ecological studies have been done in widely scattered often isolated fields. Any global solution to the problems involved will require interdisciplinary cooperation in the fields of pharmacology and toxicology, agriculture, biology, pathology, pharmacy, industrial medicine and hygiene, clinical medicine, veterinary medicine, clinical chemistry and biochemistry. Common problems and methods from these varied disciplines have given impetus to a new specialty, environmental health sciences; it is to be hoped that this field will provide a suitable context for continued systematic work on the topic of porphyria and the environment, with emphasis on lead and polyhalogenated aromatic compounds[1]. Finally, vinyl chloride disease[2] should be mentioned: this industrial disease involves chronic liver damage, scleroderma-like cutaneous changes, thrombocytopenia, coproporphyrinuria and chronic hepatic porphyria[3].

In workers of the polyvinyl chloride (PVC) producing and processing industry a complex syndrome was observed involving skin, vessels, bones and inner organs (table 1) and which was designated "vinyl chloride disease"[2]. Hepatic lesions are the most significant disturbances in this disease[3].

TABLE 1
SYMPTOMS AND SIGNS OF VINYL CHLORIDE DISEASE[2,3]

Scleroderma-like skin changes
Raynaud's Syndrome
Clubbing of terminal phalanges
Acroosteolyses
Liver damage
Biochemical findings:
serum values of GOT, GPT, AA activities normal or slightly increased
BSP-test pathological
secondary coproporphyrinuria
chronic hepatic porphyria Type A
Histological findings:
mild hepatocellular alterations
intralobular, perisinusoidal,
portal and capsular fibrosis
Hemangioendotheliosarcoma of the liver
Esophageal varices
Splenomegaly
Thrombocytopenia
Reticulocytosis

MATERIAL AND METHODS

In order to find biochemical parameters indicating toxic effects at an early stage, we have determined porphyrin precursors and porphyrins in the urine of 40 patients from a PVC producing and processing plant. Each patient examined showed one or several of the following symptoms[3]: scleroderma-like skin changes, Raynaud's syndrome, clubbing of terminal phalanges, acroosteolyses, pathological BSP test, esophageal varices, splenomegaly, thrombocytopenia, reticulocytosis. None of the patients from the PVC processing plant showed Raynaud's syndrome, scleroderma-like skin schanges, acroosteolyses or esophageal varices. Time of exposition in the group of PVC producing workers ranged from 1.75 to 20 years, in the group of PVC processing workers from 1 to 13 years. At the time of examination 12 of the patients from the PVC production plant had had no vinyl chloride contact for a time ranging from 3 to 43 months. Alcoholism could be excluded as cause for the porphyrinopathies.

RESULTS

The results of urinary porphyrin studies are given in table 2. Nineteen patients exhibited a mild secondary coproporphyrinuria, and in eight patients a moderate secondary coproporphyrinuria was found. Excretion of δ-aminolevulinic acid was elevated in four cases with mild coproporphyrinuria. In ten patients with coproporphyrinuria, excretion of pentacarboxylicporphyrin was found to be in the upper range. In a few cases critical amounts of tricarboxylicporphyrin were found. The characteristic porphyrin pattern of chronic hepatic porphyria Type A was observed in three cases (table 2).

DISCUSSION

The most frequent symptom amongst the patients examined is the elevated coproporphyrin excretion, apart from thrombocytopenia, which can be related in this connection to toxic effects by vinyl chloride. There is no correlation between the amount of urinary porphyrin excretion and the activities of "liver enzymes" in the serum.

A better correlation is seen between the BSP test and coproporphyrinuria. In patients who showed the symptoms thrombocytopenia, splenomegaly, pathological BSP test and on whom liver biopsis were performed so far, histological examination revealed no changes which indicated toxic influence by alcohol[3].

The investigations show clearly that the majority of the patients of the PVC producing and processing plants, whose biochemical and scintigraphic findings indicated liver damage, developed secondary coproporphyrinuria sometimes with slight δ-aminolevulinic aciduria. This disturbance of porphyrin metabolism led in some cases, which present an elevated excretion of uroporphyrin and heptacarboxylicporphyrin, to chronic hepatic porphyria Type A (table 2). It can be suggested that by some toxic processes vinyl chloride diminishes the activity of hepatic coproporphyrinogen oxidase and in some (susceptible?) persons also the activity of hepatic uroporphyrinogen decarboxylase. It can not be excluded that the uroporphyrinogen decarboxylase defect is purely toxic in these workers, depending upon dose and time factors of exposure. The results allow the conclusion that the determination of urinary porphyrins might become an important

TABLE 2

URINARY PORPHYRIN AND PORPHYRIN PRECURSORS EXCRETION IN 40 PATIENTS WITH VINYL CHLORIDE DISEASE IN A PVC PRODUCING AND PROCESSING PLANT;

Groups: I no disturbance in porphyrin metabolism, II mild secondary coproporphyrinuria, III moderate secondary coproporphyrinuria, IV chronic hepatic porphyria type A. For groups I, II and III mean values are given.

Group	VC exposure (years)	ALA	PBG	Total Porphyrins	Uro	Hepta	Hexa	Penta	Copro	Tri	Proto
		μmol/24 h ($\bar{x} \pm s$)		nmol/24 h ($\bar{x} \pm s$)	(%)						
I (n=10)	1-21	25 ± 9	3.6 ± 2.1	108 ± 26	13	3	1	3	79	1	-
II (n=19)	2-18	35 ± 17	5.4 ± 2.7	186 ± 26	14	3	1	4	77	1	-
III (n= 8)	3-18	26 ± 9	5.9 ± 2.5	263 ± 53	12	2	1	4	80	2	-
IV : Patients CHP A											
1. G.K.(61)	13	28	3.5	603	18	10	1	3	67	1	<1
2. W.S.(46)	7	19	4.4	340	17	12	2	5	63	1	<1
3. K.M.(26)	3	12	2.6	382	21	9	1	4	64	1	<1
Upper normal limit		49	7.5	130							

parameter in medical monitoring and protection of vinyl chloride exposed workers.

ACKNOWLEDGEMENT

The pathobiochemical studies were supported by the Deutsche Forschungsgemeinschaft (Grant Do 134).

REFERENCES

1. Doss, M. (1977) Porphyria and environment, in: Clinical Biochemistry and Regulation of Porphyrin Metabolism. Proc. German-Brazilian Seminar on Medicine and Biomedicine (Rio de Janeiro 1975), DAAD, Bonn-Bad Godesberg, 21.
2. Veltman, G., Lange, C.E., Jühe, S., Stein, G. and Bachner, U. (1975) Ann. N.Y. Acad. Sci., 246, 6.
3. Lange, C.E., Bloch, H., Veltman, G. and Doss, M. (1976) Urinary porphyrins among PVC workers, in: Porphyrins in Human Diseases (ed. M. Doss), Karger, Basel, 352.

Chemical Porphyria in Man, J.J.T.W.A. Strik and J.H. Koeman eds.

URINARY D-GLUCARIC ACID AND URINARY TOTAL PORPHYRIN EXCRETION IN WORKERS EXPOSED TO ENDRIN

W.G. VRIJ-STANDHARDT[a], J.J.T.W.A. Strik[a], C.F. OTTEVANGER[b] and N.J. VAN SITTERT[c]

[a]Department of Toxicology, Agricultural University, Wageningen, The Netherlands

[b]Department of Occupational Health, Shell Nederland Chemie, Rotterdam, The Netherlands

[c]Shell Internationale Research Maatschappij, The Hague, The Netherlands

SUMMARY

Workers producing endrin were examined for their urinary excretion of D-glucaric acid and total porphyrin. The urines were examined after a 7 day period of working at the plant and also after a period of 3 days away from the plant.

No difference was found between the porphyrin excretion in the exposed workers after working and after a long weekend of three days, and a control group (office workers). The excretion of D-glucaric acid after working was significantly increased compared to excretion in workers after the above-mentioned long weekend, and to the control group. There was no correlation between the urinary excretion of D-glucaric acid and total porphyrins.

INTRODUCTION

Endrin is a chlorinated hydrocarbon compound, chemically related to substances known to be porphyrinogenic (HCB, PCB) in animals and man. Endrin may enhance D-glucaric acid excretion[3]. The objective of the present study was to investigate whether a relationship exists between porphyrin and D-glucaric acid excretion in workers exposed to endrin.

D-glucaric acid

Most lipid soluble xenobiotics are transformed in the liver into more polar substances. This happens by oxidation, reduction, hydrolysis and conjungation reactions. Conjungation reactions occur with several endogenous compounds, such as glucuronic acid, sulfate, glycine and glutathione. In most mammals, the reaction with D-glucuronic acid may be considered as a most important route of drug-metabolizing. In its activated form D-glucuronic acid is able to react with several functional groups (see fig. 2, Ottevanger and van Sittert, this volume).

Several investigators[1,2] have demonstrated that a large number of exogenous compounds which enhance the urinary excretion of L-ascorbic acid in rats, also alter the activities of the drug-metabolizing enzymes. Man and the guinea-pig, however, do not have the ability to synthetize L-ascorbic acid because of the absence of an enzyme.

Later it was observed that treatment with a large number of different chemicals may result in an enhanced excretion of D-glucaric acid. Both D-glucaric acid as well as L-ascorbic acid are products of the glucuronic acid pathway. The determination of the urinary excretion of D-glucaric acid is used as a simple non-selective test for exposure to xenobiotics[3].

Porphyrins

Porphyrins are substances with a nucleus of four pyroles, interconnected by methene bridges, and with a different number of COOH groups. Uro-, copro- and protoporphyrin (with resp. 8, 4 and 2 COOH groups) are the most common. Protoporphyrin is the direct precursor of haem. The other porphyrins are by-products in haem-synthesis in the liver. They are mainly excreted in the urine.

Liver injury, caused by foreign compounds may cause an enhanced excretion of uro- and coproporphyrins. The increased excretion of porphyrins is considered to be an early indication of a disturbance in the liver function[4].

Relation between D-glucaric acid and porphyrin excretion

In Porphyria Cutanea Tarda patients, who produce excessive amounts of porphyrins, an increased urinary excretion of both porphyrins and D-glucaric acid is observed. The amounts of the two types of compounds excreted show a positive correlation (correlation coefficient circa 0.78)[5].

The intake of alcohol aggravates the symptoms in Chronic Hepatic Porphyria patients[6]. Alcohol itself may also induce an increased excretion of D-glucaric acid[7,8]. Hexachlorobenzene and PCB's (polychlorinated biphenyls) are also known to enhance total porphyrin[9] as well as D-glucaric acid excretion[10,11] in animals.

METHODS

In preliminary studies observations were made on the daily fluctuation of the urinary excretion of total porphyrins and D-glucaric acid. During 7 consecutive days total porphyrins, D-glucaric acid and creatinine excretion were determined in the early morning urine of 5 healthy young men of the department of Toxicology.

Morning urines of workers at a plant where the insecticide Endrin is produced, were collected. The urine samples were collected on the 7th work day and after two or three days leave. Office workers at the same factory were used as controls, and from these persons morning urine was also collected. D-glucaric acid was determined according to Marsh[2].

Total porphyrin was determined according to Doss[12] and this volume. Creatinine concentration was estimated according to the method of Gorter and De Graaff[13].

RESULTS

Preliminary studies

The mean porphyrin concentration of all 35 urine samples of the five healthy young men was 48.3 ± 23.3 (s.d.) μg/l.

The mean creatinine concentration was 16.6 ± 6.1 (s.d.) mmol/l. A statistically significant correlation was found between the porphyrin and creatinine excretion (see fig. 1) (correlation coefficient: 0.59).

The mean D-glucaric acid concentration of these urines was 18.3 ± 12.9 (s.d.) mg/l. Again a statistically significant, positive correlation with creatinine was found (see fig. 2). No correlationship was found between D-glucaric acid and porphyrin excretion (t = 0.11).

Studies in the endrin plant

Fig. 3 shows a scatter diagram of the urinary D-glucaric acid excretion (in mg/mmol creatinine) and the urinary total porphyrin excretion (in μg/mmol creatinine) in the endrin workers and in the control group.

D-glucaric acid. After seven days work at the endrin plant D-glucaric acid excretion in the workers was significantly higher than in those who had just had three days leave; it was also significantly higher than in the control group. The mean value after seven days of shift work amounted to 1.8 ± 0.7 mg/mmol creatinine, while the mean values after the long weekend, and the control values amounted to 1.4 ± 0.6 mg/mmol creatinine and 1.3 ± 0.7 mg/mmol creatinine respectively.

Porphyrins. No significant difference was found in urinary porphyrin excretion between the workers after seven days of shift work (mean 3.7 ± 3.3 μg/mmol creatinine) and those returning from 3 days leave (mean 4.0 ± 4.1 μg/mmol creatinine).

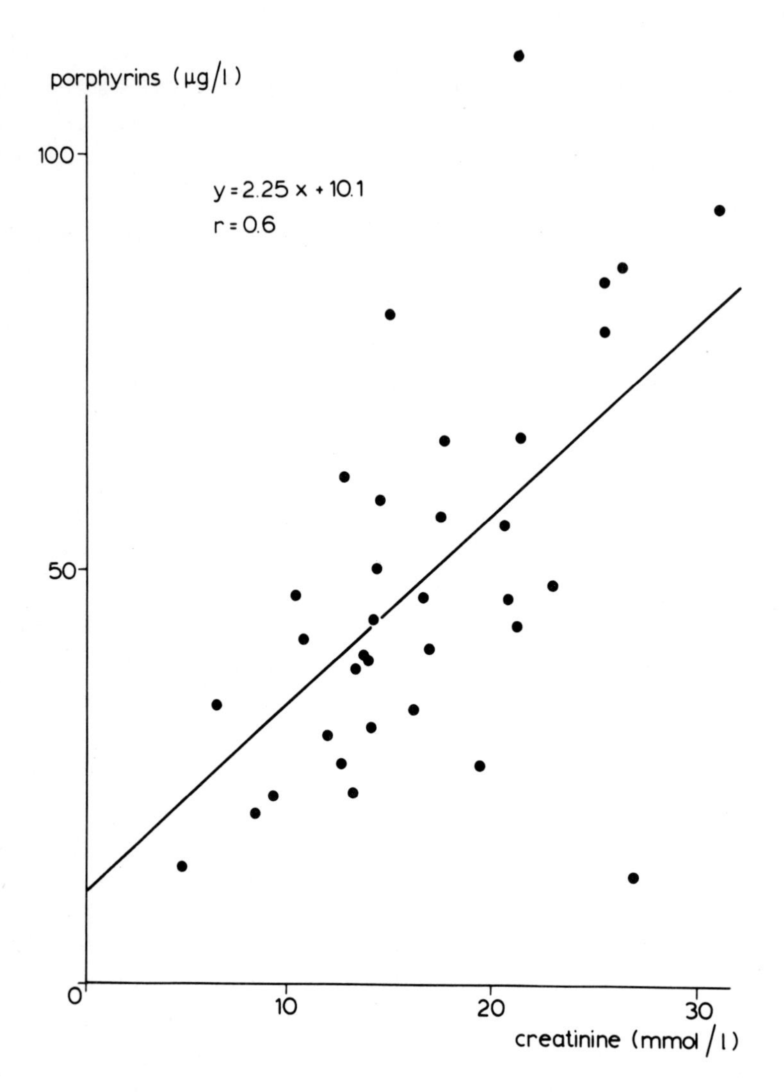

Fig. 1. The urinary porphyrin and creatinine concentration of normal persons.

Correlation D-glucaric acid - porphyrins

No correlation was found between the D-glucaric acid and the total porphyrin values of the urine samples; either within the individual groups, or in the three groups taken together (see fig. 4). The correlation coefficient r was 0.10. The urine samples which had either increased glucaric acid levels (> 1.8), or increased porphyrin values (> 6.0), or borh, did not show any correlation either.

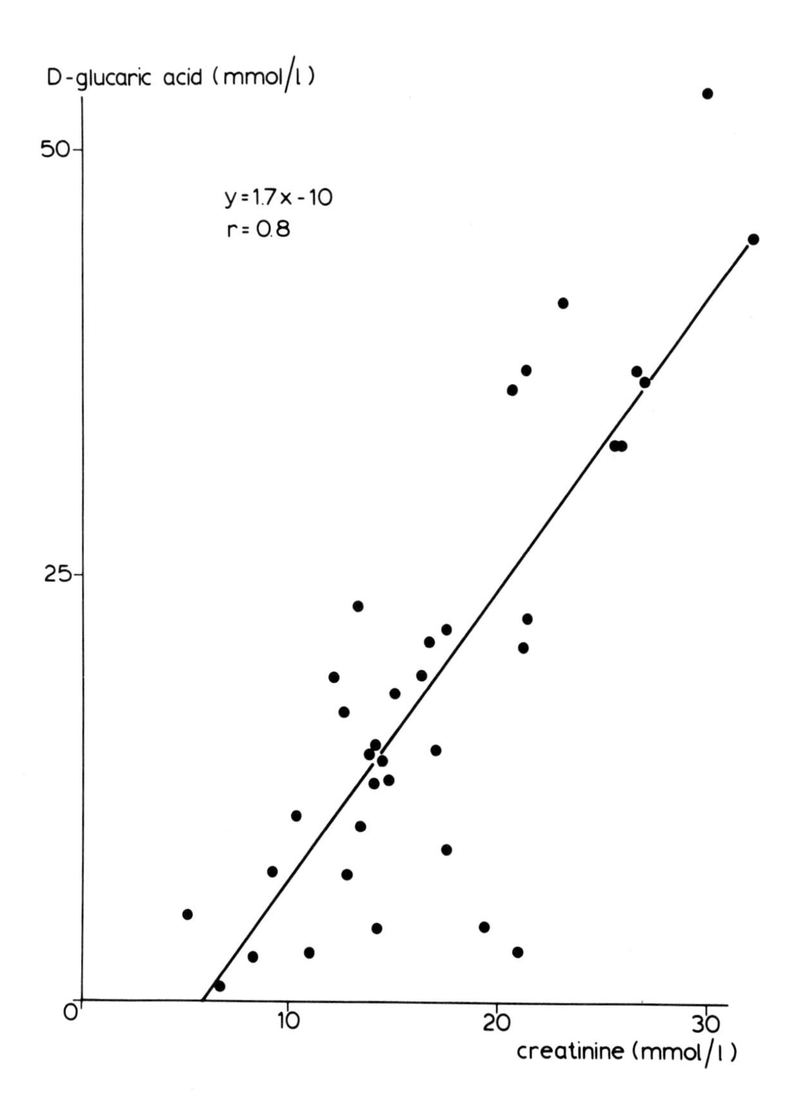

Fig. 2. The urinary D-glucaric acid and creatinine concentration of normal persons.

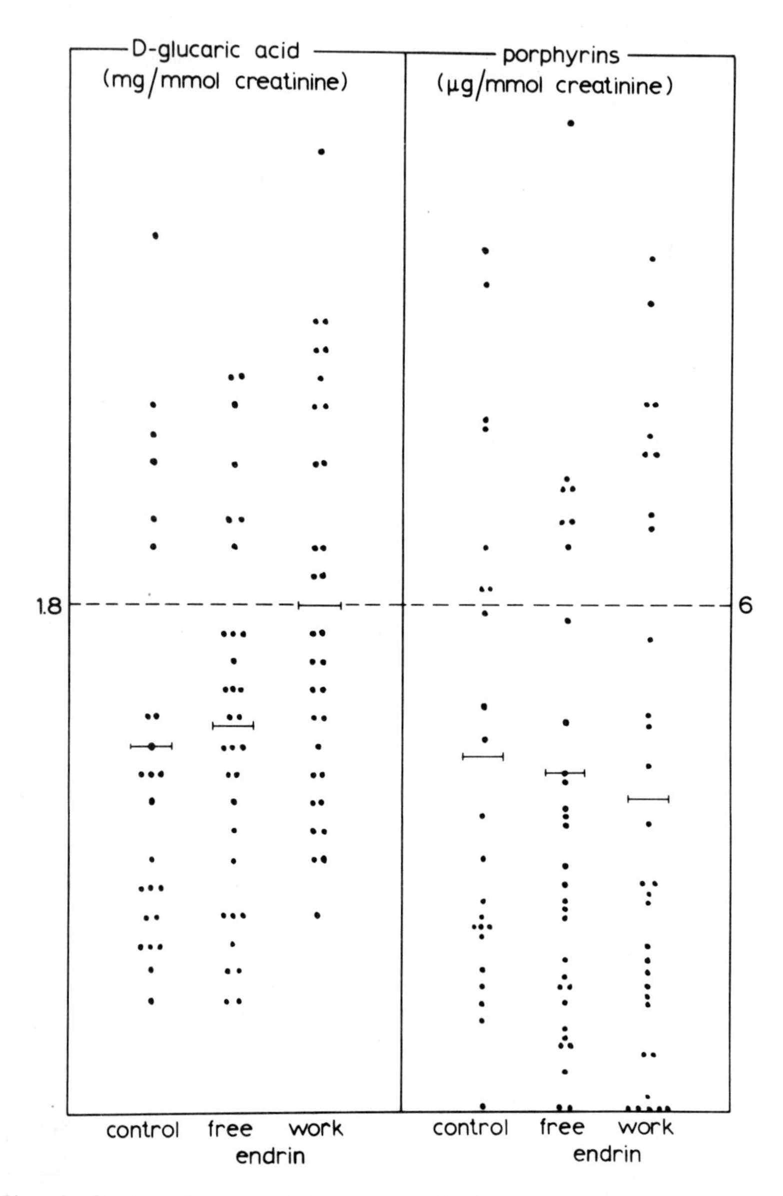

Fig. 3. Scatterdiagram of the urinary D-glucaric acid and the total porphyrin excretion in the endrin-workers after seven days work and after three days leave, and in the control group.

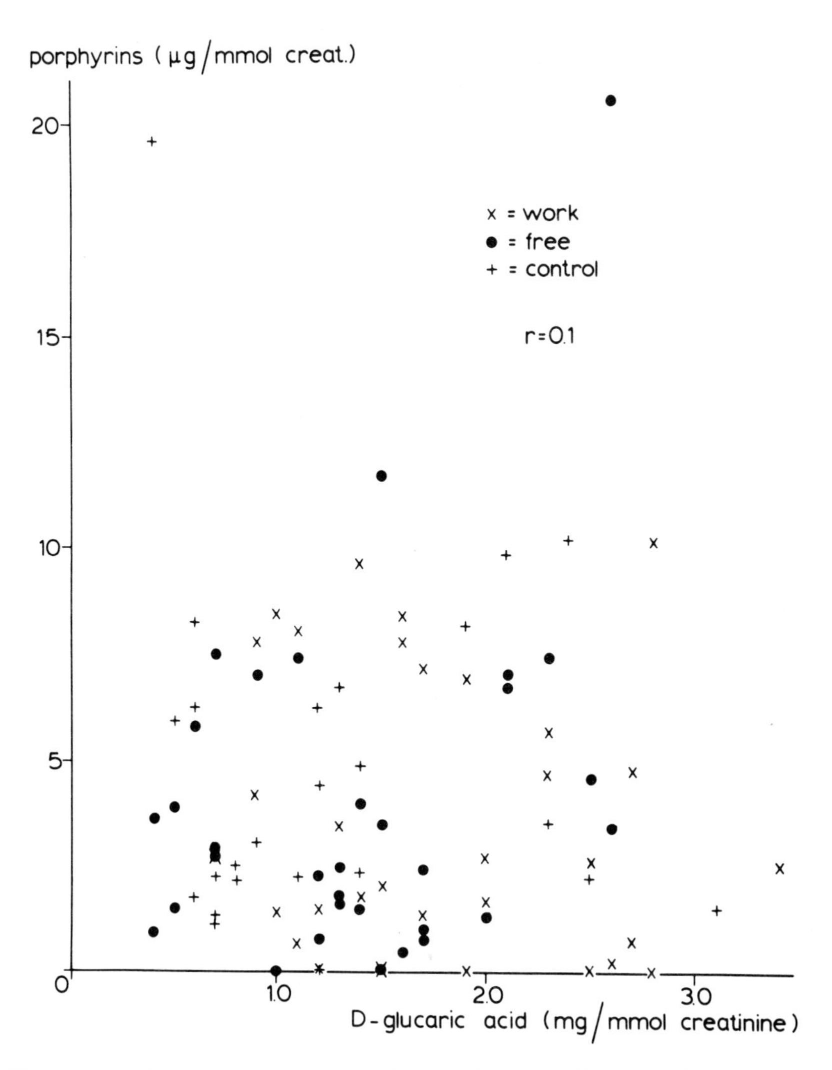

Fig. 4. The D-glucaric acid and the total porphyrin excretion in endrin-plant workers (both after work and after leave) and controls.

DISCUSSION

Both the porphyrin as well as the D-glucaric acid concentration could be correlated with the creatinine contents of the urine samples. These correlations could be expected since highly concentrated urine having a high creatinine concentration may contain more porphyrins and D-glucaric acid than diluted urine, low in creatinine content.

Therefore if one wants to assess whether or not there is a correlation between D-glucaric acid and total porphyrins in urine the degree of dilution should be taken into consideration and this can largely be achieved by expressing the levels relative to the creatinine content of the samples. The results of the preliminary studies demonstrate that expressing D-glucaric acid and porphyrins per mmol creatinine is more correct than expressing them per liter of urine.

No correlation between the urinary D-glucaric acid and porphyrin excretion in workers with exposure to endrin was found.

The D-glucaric acid levels of the workers are increased after working 7 days at the plant, the porphyrin levels are not increased.

The cases, mentioned in the literature, showing increased excretion of porphyrin as well as D-glucaric acid are all alcoholic and PCT-patients. A possible explanation for this could be: that increased excretion of D-glucaric acid is an indication of the reaction of the liver caused by the xenobiotic stimulation of hepatic drug-metabolism. Increased porphyrin excretion is an indication of dysfunctioning of the liver. A subtoxic exposure to endrin causes only a stimulation of the drug-metabolism, and so an increased D-glucaric acid excretion. In the alcoholic and PCT-patients liver dysfunction is involved.

It is also likely that endrin is not porphyrinogenic at all. Experiments with Japanese quail support this hypothesis[14].

In conclusion, determination of D-glucaric acid excretion is a useful test for exposure to endrin; determination of porphyrin excretion is not.

ACKNOWLEDGEMENTS

The authors are very grateful for the technical assistance provided by the Technological Department, the drawnings by C. Rijpma and M. Schimmel and the photography by A. van Baaren of the Biotechnion of the Agricultural University at Wageningen. Typing of the manuscript by Miss G. van Steenbergen and Mrs. L. Muller Kobold - de Lagh is greatly appreciated. We are indebted to the laboratory assistance of Miss E.G.M. Harmsen.

REFERENCES

1. Longecker, H.E., Fricke, H.H. and King, C.G. (1940) J. Biol. Chem. 135, 497.
2. Marsh, C.A. (1963) Biochem. J. 87, 82.
3. Notten, W.R.F. (1975) Alterations in the D-glucaric acid pathway and drug metabolism by exogenous compounds, Thesis, Stichting Studentenpers, Nijmegen.
4. Lange, C.E., Bloch, H., Veltman, G. and Doss, M. (1976) Urinary porphyrins among PVC workers. In: Porphyrins in human diseases. Proc. of the 1st Int. Porphyrin Meeting, Karger, Basel, 352.
5. Budillon, G., Ayala, F., Carella, M. and Mazzacca, G. (1975) Digestion, 12, 292.
6. Doss, M., Nawrocki, P., Schmidt, A., Strohmeyer, G., Egbring, R., Schmipff, G., Dölle, W. and Korb, G. (1971) Dtsch. med. Wschr. 96, 1229.
7. Spencer-Peet, J., Wood, D.C.F., Glatt, M.M. and Wiseman, S.M. (1975) Brit. J. Addict. 70, 359.
8. Mezey, E. (1976) Res. Comm. in Chem. Path. and Pharmacol. 15, 735.
9. Strik, J.J.T.W.A. (1978) J. Clin. Chem. Clin. Biochem. 16, 54.
10. Lissner, R., Görtz, G., Eichenauer, M.G. and Ippen, H. (1975) Biochem. Pharmacol. 24, 17.
11. Notten, W.R.F. and Henderson, P.T. (1977) Int. Arch. Occup. Environ. Hlth. 38, 209.
12. Doss, M. and Schmidt, A. (1971) Z. klin. Chem. u. klin. Biochem. 9, 415.
13. Gorter, E. and Graaff, W.C. de (1955) Klinische Diagnostiek, 7th ed. Stenfert & Kroese, Leiden, 440.
14. Wiel-Wetzels, W.A.M. van der and Strik, J.J.T.W.A. (1978) Porfyrines in urine na blootstelling aan endrin. Internal report, Department of Toxicology, Wageningen.

Chemical Porphyria in Man, J.J.T.W.A. Strik and J.H. Koeman eds.

RELATION BETWEEN ANTI-12-HYDROXY-ENDRIN EXCRETION AND ENZYME INDUCTION IN WORKERS INVOLVED IN THE MANUFACTURE OF ENDRIN

C.F. OTTEVANGER[a] and N.J. VAN SITTERT[b]

[a]Department of Occupational Health, Shell Nederland Chemie, Rotterdam, The Netherlands

[b]Shell Internationale Research Maatschappij, The Hague, The Netherlands.

SUMMARY

The relationship between endrin exposure and enzyme-induction has been studied in workers of the endrin manufacturing plant in Pernis. Exposure was measured indirectly by the determination of the endrin metabolite anti-12-hydroxy-endrin in urine. Enzyme induction was measured by assessing urinary D-glucaric acid levels.

It has been demonstrated that urinary D-glucaric acid levels at the end of a 7 days' shift were higher in a group of endrin workers than in a control group, indicating that exposure to endrin might produce enzyme induction. However, it has been shown that this effect is reversible, as urinary D-glucaric acid levels were within the normal range after a six week period of shut-down of the plant.

A urinary level of anti-12-hydroxy-endrin of 0.130 µg/gram creatinine was suggested as a threshold exposure level, below which enzyme induction is not produced.

INTRODUCTION

The workers in the Endrin-plant at Pernis have been under close medical supervision for many years. Jager[4] already reported in 1970 on the health of these workers and stated that the occurrence of enzyme induction is an early effect of organochlorine insecticides.

Enzyme induction is defined as enhancement of activity of microsomal enzymes. These processes occur particularly in the liver. Therefore various liver enzyme parameters in the group of Endrin workers were discussed.

- Serum alkaline phosphatase levels, SGOT and LDH levels were normal.
- The SGPT was slightly but not significantly higher than in the other groups exposed to the organo-chlorines: Dieldrin and Aldrin.
- The para para' DDE levels in blood were considered to be a good indicator

for enzyme induction. DDT is not handled in our pesticide plants and the concentrations of its metabolite pp' DDE in the blood should therefore be no different from those in the general population. It became apparent that pp' DDE levels were lower in the Endrin and former Endrin workers group (for further reference see Jager[4], Hunter and Robinson[3]).

Measurement of urinary excretion of 6-β-hydroxycortisol (a metabolite of cortisol) was also carried out in 8 Endrin workers.

The ratios of 6-β-hydrocortisol/17-OH-corticosteroids in the urine were higher in comparison to the other groups, confirming the possibility of enzyme induction following Endrin exposure as demonstrated earlier with pp' DDE levels in blood. Jager's[4] conclusion was that exposure in the Endrin plant results in enzyme induction.

However, no conclusion about the level of industrial exposure to Endrin could (and can) be drawn. A more reliable indicator of exposure might be the measurement of Endrin or its metabolites in body fluids, as already mentioned by Jager[4] in 1970.

Because Endrin could not be detected in blood (due to its rapid metabolism) - limit of detection 0.005 ppm - the only chance of determining exposure would now be the investigation of the occurrence of metabolites in urine or faeces, the latter being impractical in occupational medicine.

Baldwin and Hutson[1] demonstrated that Endrin metabolites were present in human urine and in faeces. The major metabolite of Endrin in the urine was found to be anti-12-OH-Endrin (fig. 1).

Baldwin[2] further developed a gas chromatographic method for quantitative measurement of the metabolite. Concentrations up to 0.13 μg/ml were found in the urine of Endrin workers.

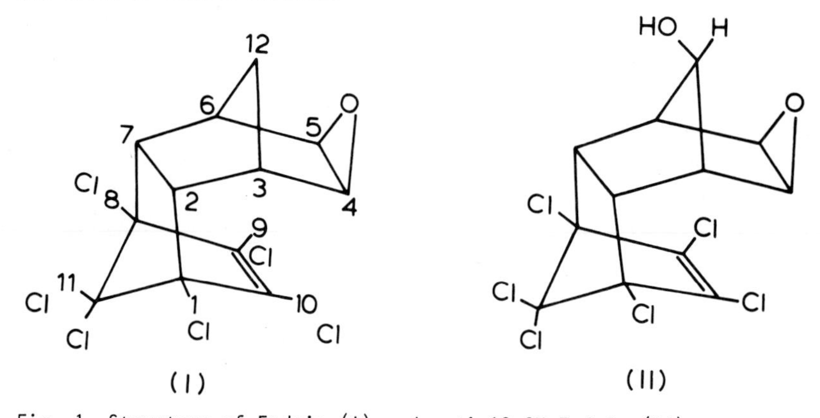

Fig. 1. Structure of Endrin (I) and anti-12-OH-Endrin (II).

In our preliminary study the urinary levels of this metabolite were determined in a number of Endrin workers.

In addition, it was felt to be of interest to continue the previously mentioned enzyme induction studies, because enzyme induction is a biological effect which might be undesirable.

Determination of the 6-β-hydroxy-cortisol is not practical for routine use, and pp' DDE is a not very specific and therefore less suitable indicator for enzyme induction. The D-glucaric acid concentration in the urine[5] is currently considered to be a better test for enzyme induction.

D-glucaric acid is formed as a by-product in the glucuronic acid cycle (fig. 2).

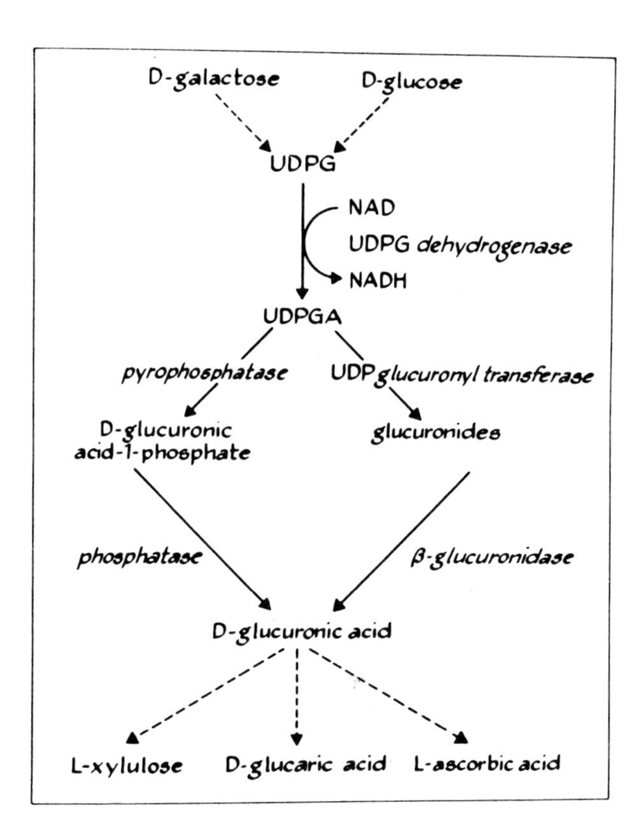

Fig. 2. The D-glucuronic acid pathway. (Reproduced with permission of W.R.F. Notten).

RESULTS

29 Endrin workers were examined three times. The first examination was done after a 6 weeks period of shut-down and maintenance of the plant; in this period exposure is considered to be very low. The second examination was done after 7 days of exposure during operation of the plant. This group was again examined after a period of 3 days in which no exposure could take place (long week-end).

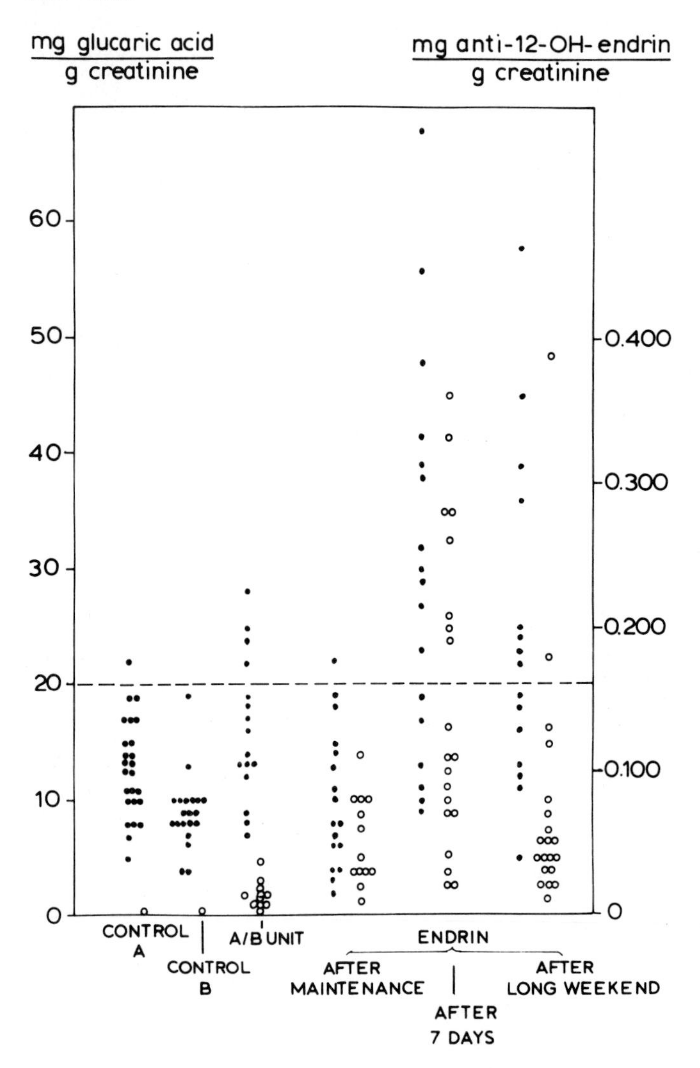

Fig. 3. D-glucaric acid and anti-12-OH-Endrin levels in urine of endrin workers after maintenance, operation and long week-end and control (A and B).

Both the anti-12-OH-Endrin and D-glucaric acid urinary levels were determined and related to the urinary creatinine concentration (the latter in order to eliminate the factor of differences in diuresis).

In control groups A and B (not exposed to Endrin) anti-12-OH-Endrin levels in the urine were below the limit of detection (<0.002 ppm). D-glucaric acid levels in these groups were in good agreement with the concentrations found by Notten[5]. These workers considered a value of D-glucaric acid of 20 mg per gram creatinine to be normal. Levels exceeding this value were considered to be an indication of enzyme induction (fig. 3).

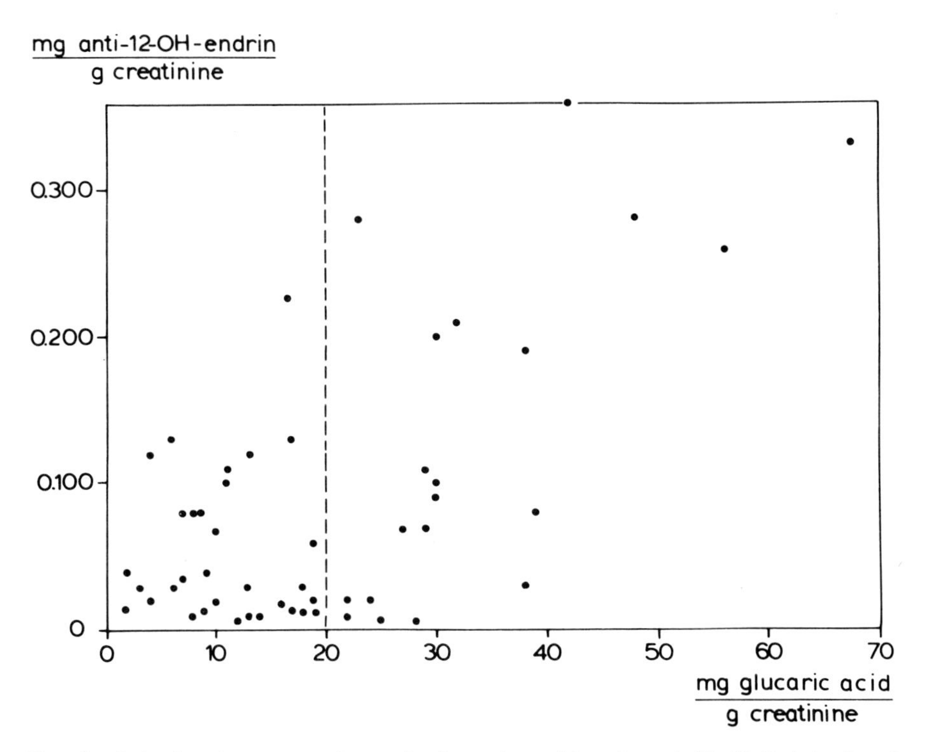

Fig. 4. Relation between urinary D-glucaric acid and anti-12-OH-Endrin levels of Endrin manufacturing workers.

The group of workers examined directly before the start-up operations (after maintenance) of the plant had anti-12-OH-Endrin levels in the urine generally below 0.100 mg per gram creatinine adjusted to creatinine. The D-glucaric acid levels were in the same range as in the control groups.

After 7 days of exposure the metabolite levels increased (up to 0.360 mg per gram creatinine) accompanied by a sharp rise in D-glucaric acid levels. The latter might indicate that an enzyme induction had already occurred following 7 days' exposure to Endrin. After the long weekend the anti-12-OH-Endrin concentrations decreased again very clearly, but D-glucaric acid levels remained on a higher level when compared to normal values. However, the fact that the workers in this group had D-glucaric acid levels within the normal range after the six week's maintenance period indicates that enzyme induction in Endrin workers is reversible as we may assume that previous exposure levels were similar to the exposure during the study.

In another group of pesticide workers (formulators and cleaners of the A/B Unit) the metabolite levels were clearly below 0.100 mg per gram creatinine while D-glucaric acid levels were in the normal range (fig. 3).

DISCUSSION

As regards the dose-effect relationship between Endrin exposure (or the excretion of the Endrin metabolite) and urinary D-glucaric acid levels, our preliminary results show a correlation between these parameters when urine samples are taken at the end of a seven days working week (fig. 4). About 75% of measurements with urinary anti-12-OH-Endrin levels below 0.130 mg per gram creatinine corresponded with D-glucaric acid concentrations which were in the normal range.

This suggests that this level is a threshold value below which enzyme induction does not occur.

However, increased D-glucaric acid levels were occasionally found when anti-12-OH-Endrin levels were below 0.130 mg per gram creatinine. This, assuming a threshold value of 0.130 mg per gram creatinine (at the end of a working week), would represent false positives with respect to enzyme induction. This might be explained as follows:

a) Endrin exposure did not occur recently; anti-12-OH-Endrin levels might return to low levels more readily than D-glucaric acid levels.
b) Other compounds e.g. alcohol and drugs may produce enzyme induction.
c) Susceptibility to enzyme inducers may vary between individuals.

The quantitative relationship between Endrin exposure and levels of the metabolite in urine are not yet known. The quantitative relationship between the level of exposure to Endrin and enzyme induction is still unknown. Further studies to confirm the level of Endrin exposure at which no effect on enzyme induction can be demonstrated, and to correlate excretion with Endrin exposure, are required.

REFERENCES

1. Baldwin, M.K. and Hutson, D.H. (1973) Non-published results, Shell Toxicology, Laboratory Tunstall, U.K.

2. Baldwin, M.K. (1973) Non-published results, Shell Toxicology Laboratory Tunstall, U.K.

3. Hunter, J. and Robinson, J. (1972) Nature, 237, 399.

4. Jager, K.W. (1970) Aldrin, Dieldrin, Endrin and Telodrin, an epidemiological and toxicological study of long-term occupational exposure, Thesis, Elsevier Publishing Company, Amsterdam.

5. Notten, W.R.F. (1975) Alterations in the D-glucuronic acid pathway and drug metabolism by exogenous compounds, D-glucaric acid levels as an indication of exposure to xenobiotics, Thesis, Stichting Studentenpers, Nijmegen.

Chemical Porphyria in Man, J.J.T.W.A. Strik and J.H. Koeman eds.

URINARY THIOETHER AND D-GLUCARIC ACID EXCRETION AFTER INDUSTRIAL EXPOSURE TO PESTICIDES

F. SEUTTER-BERLAGE, M.A.P. WAGENAARS-ZEGERS, J.M.T. HOOG ANTINK and H.A.F.M. CUSTERS

Department of Pharmacology, University of Nijmegen and Medical Department, Luxan B.V., Elst, The Netherlands.

SUMMARY

The urinary thioether and D-glucaric acid excretion of a group of workers in the production department of a pesticide factory were compared with those of a group of workers in the administration department in the same factory. Pesticide exposed workers had a significantly higher thioether excretion than administrative workers. In addition it appeared that both in the pesticide exposed group or the administrative group smokers showed a significantly higher thioether excretion than non-smokers.

In either the smoking or the non-smoking group the groups exposed to pesticides showed a significantly higher D-glucaric acid excretion than their colleagues in the administration department. The glucaric acid test showed no effect due to smoking.

INTRODUCTION

Mercapturic acid derivatives or other thioethers can be considered as the end products of the metabolic detoxification of alkylating agents[1,2,3].

To detect expositions to compounds with potentially alkylating properties we developed a thioether test in urine[4]. In this investigation we applied this test to urine samples from a number of workers from a factory producing agricultural chemicals. The factory does not synthesize chemicals but powders, formulates, mixes, packs and sells the final product to the users. The production includes insecticides (ranging from halogenated hydrocarbons to organic phosphor compounds), fungicides, herbicides, fertilisers, etc.

Since it has been reported in the literature[5,6] that an enhanced urinary excretion of D-glucaric acid might be expected with occupational exposure to certain chlorinated hydrocarbon pesticides, we also determined the D-glucaric acid excretion.

We compared samples of urine from a group of workers from the pesticide producing department with those from a group of administrative workers in the same factory in order to eliminate environmental expositions other than caused

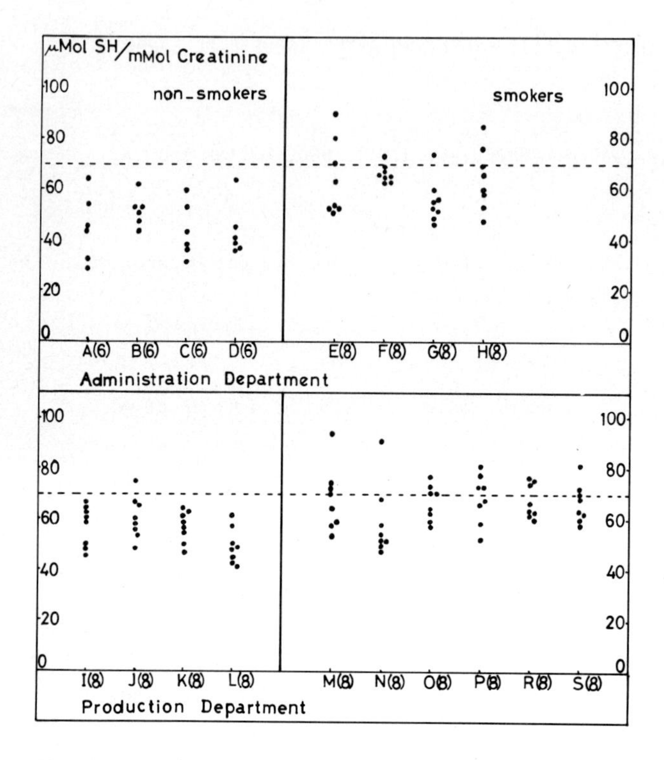

Fig. 1. Thioether excretion in smokers versus non-smokers in the administration department and in the production department of a pesticide factory. The dotted line indicates the average background value + S.D.

by the manufacturing process, to see whether, regardless of extensive safety precautions, a difference in work exposition could be shown.

MATERIAL AND METHODS

The workers (8 in the administration department, 10 in the production department), volunteering in this investigation were all male and not under any medication.

Urine samples were collected from each individual once a week at the end of a working day. The number of samples, taken during the same period, from each individual varied from 6 to 8. The samples were inmmediately frozen and stored at -10 °C until analyzed.

The thioether test was performed after alkaline hydrolysis of the thioethers to thiols as published earlier[4]. D-glucaric acid was estimated according to the method of Marsh[7] from the inhibitory effect of D-glucaro-1,4,-lacton (to which it is converted by heating at pH 2) on β-glucuronidase.

Creatinine concentrations in the urine were estimated according to Gorter and de Graaff[8].

The results were expressed as μMol SH/mMol creatinine or μMol D-glucaric acid/mMol creatinine.

Statistical analyses were performed by the Wilcoxon two sample test[9].

RESULTS

Individuals were classified according to working in the administration department or in the production department, and according to smoking or non-smoking. Wilcoxon's two sample test was applied to the four subgroups.

The smokers both in the administration department (n=4) or in the production department (n=6) had a significantly higher thioether excretion than the non-smokers ($p<0.01$) (fig. 1). Those exposed to pesticides either in the smoking or in the non-smoking group (n=4) had a significantly higher D-glucaric acid excretion than their colleagues in the administration department ($p<0.01$) (fig. 2).

The D-glucaric acid test showed no significant effect due to smoking. The group of smoking administration workers had a significantly higher thioether excretion than the non-smoking production group ($p<0.02$) (fig. 1), but a significantly lower D-glucaric acid excretion ($p<0.01$) (fig. 2).

Assuming a normal distribution of the parameter values, correlation coefficients were calculated[9] for the thioether and creatinine excretions for each individual. They ranged from 0.743 for the man with the highest thioether excretion to 0.826 for the one with a low thioether excretion. In all cases correlations were significantly positive ($p<0.05$).

DISCUSSION

Upon questioning it appeared that the man from the administration department who showed a very high thioether excretion, even higher than most of his colleagues in the production department, was a heavy smoker. We accordingly subdivided the groups into smokers and non-smokers.

It is well-known that in smoking tobacco and exposition to aromatic hydrocarbons occurs[10]. The toxicity, mutagenicity, or carcinogenicity of these polycyclic aromatic hydrocarbons depends on their metabolism by the non-specific microsomal mono-oxygenases to reactive electrophilic intermediates (epoxides)[11]. Detoxification of these compounds occurs, at least in part, through conjugation

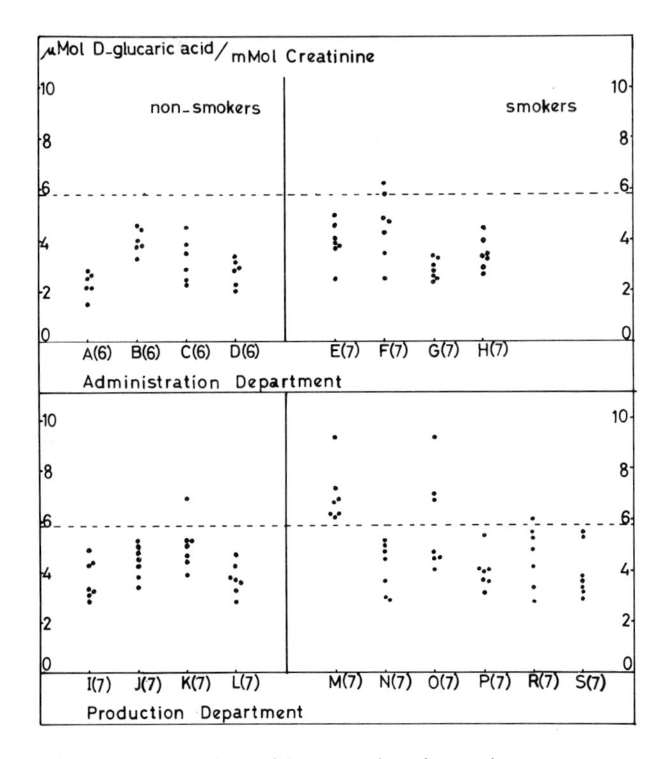

Fig. 2. D-glucaric acid excretion in smokers versus non-smokers in the administration department and in the production department of a pesticide factory. The dotted line indicates the average background value + S.D.

with endogenous glutathione, resulting in urinary excretion of thioethers[1,2,3]. In accordance with this it has to be expected that smokers might show higher thioether excretions than their non-smoking colleagues, working under the same conditions of other expositions. However, in investigating pesticide expositions, resulting in similar detoxifications[3,12] we did not expect the effect of smoking to be stronger than that of the pesticides.

We did expect an stimulatory effect of the pesticide exposition on the D-glucaric acid excretion[5,6]. This was in accordance with our findings, however, the values found did not surpass the normal values reported in literature[5].

Obviously the safety precautions in the factory were quite efficient against expositions. This is in accordance with the strongly positive correlations we found between urinary creatinine and thioether excretion. In serious intoxications this correlation is mostly lost, whereas in individuals only slightly exposed to alkylating agents the thioether excretion depends for the greater part on normal protein breakdown.

ACKNOWLEDGEMENT

We are indebted to Drs. E. Seutter, Department of Dermatology, University of Nijmegen, for his valuable advice in this study.

REFERENCES

1. Wood, J.L. (1970) Biochemistry of mercapturic acid formation, in: Metabolic conjugation and metabolic hydrolysis, W.H. Fishman (ed.), Academy Press, Vol. II, p. 261.
2. Boyland, E. (1971) Mercapturic acid conjugation, in: Handbook of Exp. Pharmac., B.B. Brodie and J.R. Gilette, (eds.), Springer Verlag, Berlin, Vol. 28, II, p. 584.
3. Chasseaud, L.F. (1976) in: Glutathione: Metabolism and Function, I.M. Arias and W.B. Jakoby (eds.), Raven Press, New York, p. 77.
4. Seutter-Berlage, F., Dorp, H.L. van, Kosse, H.G.J. and Henderson, P.T. (1977) Int. Arch. Occup. Environm. Hlth., 39, 45.
5. Nötten, W.R.F. and Henderson, P.T. (1977) Int. Arch. Occup. Environm. Hlth., 38, 209.
6. Hunter, J., Maxwell, J.D., Stewart, D.A., Williams, R., Robinson, J. and Richardson, A. (1972) Nature, 237, 399.
7. Marsh, C.A. (1963) Biochem. J., 86, 77.
8. Gorter, E. and Graaff, W.C. de (1955) in: Klinische diagnostiek, 7th ed., Stenfert & Kroese, Leiden, p. 440.
9. Diem, K. (1960) in: Wissenschaftliche Tabellen, Geigy, Basel, p. 170-10, p. 170-23.
10. Severson, R.F., Smook, M.E., Bigman, H.C., Chortyk, O.T. and Akin, F.J. (1974) in: Carcinogenesis, R.I. Freudenthal and P.W. Jones (eds.), Acad. Press, New York, Vol. I, p. 253.
11. Sims, P. and Grover, P.L. (1974) in: Advances in Cancer Research, G. Klein and S. Weinhouse (eds.), Acad. Press, New York, Vol. 20, p. 166.
12. Hutson, D.H. (1977) Chem.-Biol. Interactions, 16, 315.

EXPERIMENTAL STUDIES

Chemical Porphyria in Man, J.J.T.W.A. Strik and J.H. Koeman eds.

EXPERIMENTAL ERYTHROPOIETIC PROTOPORPHYRIA

FLUORESCING CELLS IN THE BLOOD AND THE BONE MARROW OF MICE AND JAPANESE QUAIL AFTER FEEDING 3,5-DIETHOXYCARBONYL-1,4-DIHYDROCOLLODINE, GRISEOFULVIN OR LEAD

R.A. WOUTERSEN, C.W.M. van HOLSTEIJN and J.G. WIT

Institute of Veterinary Pharmacology and Toxicology, State University, Biltstraat 172, Utrecht, The Netherlands.

SUMMARY

Japanese quail and mice were fed a standard powdered diet containing lead, 3,5-diethoxycarbonyl-1,4-dihydrocollidine (DDC) or griseofulvin (GF). Administration of DDC or GF to mice primarily induced a hepatic porphyria. The blood plasma contained detectable amounts of protoporphyrin (PP) before porphyrins were detectable in the blood cells or bone marrow cells. Our results suggest that the accumulated PP in the fluorescing cells is hepatic, not erythropoietic in origin. An interference of DDC or GF with the haem synthesis in the bone marrow seems unlikely but is not excluded. Our experiments with bone marrow cell cultures might provide us with more information about this aspect. The results of some preliminary in vitro experiments with the bone marrow cells and blood cells of Japanese quail are reported.

Lead caused the occurrence within 24 hrs of many fluorescing cells in the blood of quail. An accumulation of PP in the plasma was hardly detectable. In mice, only very few fluorescing cells were observed in the blood and the bone marrow after administration of dietary lead, however, the porphyrin concentration in the spleen was highly elevated (mainly CP). We conclude that the presence of fluorescing cells in the blood of lead poisoned quail is due to the interference by lead of the haem synthesis in circulating immature erythrocytes and bone marrow cells causing an accumulation of PP.

Photohaemolysis experiments indicated a difference between the fluorescing cells in the blood of DDC or GF treated mice, lead poisoned quail or EPP-patients, with regard to their sensitivity to irradiation with 410 nm light. Blood cells from DDC or GF treated mice were very sensitive to irradiation, probably due to the accumulation of PP, mainly in the cell membranes.

The different origins of the fluorescing cells in DDC/GF treated mice and lead posoned quail, and the absence of fluorescing cells in the blood of hexachlorobenzene or hexabromobiphenyl intoxicated quail, is discussed.

INTRODUCTION

Since the early 1880's it has been known that lead has an effect on haemopoiesis and disturbs porphyrin metabolism (Stokvis[20], Behrend[2]). Salomon et al.[15] and deMatteis et al.[10] discovered the porphyrinogenic activity of DDC and GF respectively. The appearance of fluorescing cells or fluorescing plasma in the blood of animals treated with these agents has been mentioned (Salomon[15], Fileder[6], Roscoe[14]).

The effects of lead, DDC and GF on haem synthesis in bone marrow cells are unknown. Schmid[17] observed the accumulation of UP I in the bone marrow of rabbits exposed to lead, phenylhydrazine and light. It is known that nucleated avian erythrocytes synthesize determinable amounts of porphyrins in vitro[16]. Porphyrin biosynthesis in bone marrow cell cultures has never been studied.

Administration of GF to mice is used as a model for human EPP[8]. Protoporphyrin strongly absorbs 410 nm light; EPP-patients[7] and GF treated mice[9] show photosensitivity when they are exposed to 410 nm light. Allen et al.[1] reported three patients with blistering of the skin possible caused by lead intoxication. However, in lead toxicity cutaneous photosensitivity is usually absent[12].

We investigated the role of the bone marrow as a possible important source of fluorescing cells in cases of toxic porphyria. The fluorescing cells in the blood of lead poisoned quail or DDC/GF treated mice were compared with the fluorescing cells in the blood of EPP-patients with regard to their sensitivity to 410 nm light irradiation.

We used birds as well as mice because of their highly active erythropoiesis and the presence of mitochondria in avian erythrocytes. Furthermore we used birds because little is known about the influence of foreign chemicals on biochemical processes in these animals.

MATERIALS AND METHODS

Japanese quail and mice were fed a standard powdered diet containing:

quail: 2.5 % (w/w) GF (provided by I.C.I. Holland B.V.);
0.5 % (w/w) DDC (Eastman Kodak N.V., Rochester N.Y. 14650, USA)
or 625, 1250, 2500, 5000 ppm lead.

mice : 2.5 % (w/w) GF; 0.1 % (w/w) DDC or 5000 ppm lead.

Blood samples of EPP-patients were provided by dr. H. Baart de la Faille, Dermatological Division of the Academic Hospital, Utrecht.

The percentage of fluorescing cells in the blood and the bone marrow and the percentage of the liver, spleen- and kidney tissue that showed porphyrin fluorescence was determined using a Zeiss fluorescence microscope (with a HBO W/4 Mercury lamp), fitted with excitation filters BG 38/25 and BG 12/4, and barrier filters 53 and 33. All microscopic observations were carried out in a dark room with scarce red light.

The porphyrin patterns in the different blood fractions and the bone marrow fraction were analysed according the method described by Doss[5]. For these TLC-observations DC Fertigplatten, Kieselgel F254, Merck, W. Germany, were used. The solvent system chloroform-methanol (94:6, v/v, 3-4 cm) was used to form a base line[4]. The PP-concentration was estimated semi-quantitatively for each fraction.

For the quantitative PP determinations the PP was extracted from the blood cells and blood plasma according the method described by Piomelli[11] but with some modifications. The porphyrins were extracted from 20 µl plasma or blood with ethylacetate/acetic acid (4:1). The PP concentration in 1.5M HCl was determined using a Beckman, model 25 spectrophotometer with cuvettes with a light path of 2 cm. The calculations were carried out as described by Cripps et al.[3].

Photohaemolysis experiments

Irradiation experiments were carried out using a modification of the method described by Schothorst[18]. Freshly heparinized blood was centrifuged (3000 rpm, 5') and washed three times in a buffered isotonic NaCl solution. 200 µl of the erythrocyte suspension (Ht 35-40 %) was added to 4.2 ml saline or to 4.0 ml saline with 200 µl plasma. A Philips HPW (125 Watt) mercury lamp was used as light source.

In vitro experiments

Bone marrow cells (BMC's) were collected by flushing 2 femurs with saline. Bone marrow cell suspensions consisted of 2.10^6 cells/ml; 20 % serum (foetal bovine serum; Flow). 1 mM d-amino-levulinic acid (Sigma); 0.7 mM (NH_4) Fe $(SO_4)_2.6H_2O$. The blood cell cultures contained 2 mM ALA; 10 % serum and 5.10^6 cells/ml. The cell suspensions were incubated in Minimal Essential Medium (Flow) in Costar dishes in a CO_2-incubator for 24 or 48 hrs. 5 ml of the cell suspension was used to determine the porphyrin pattern in the cell fraction and in the medium by TLC. Quantitative porphyrin determinations were carried out (with 2 ml of a BMV- or blood cell suspension) according the method described by Cripps et al.[3].

The amount of porphyrins in the cells and in the medium were determined spectrophotometrically. Up was separated from the other porphyrins (PP) by washing with a saturated sodium acetate solution.

RESULTS

Administration of DDC or GF evoked an enhanced synthesis and accumulation of porphyrins in the liver of mice within 24 hrs, causing a strong fluorescence of the liver tissue. The plasma contained elevated protoporphyrin concentrations at this time, while the blood cells and the bone marrow cells did not contain detectable amounts of porphyrins (table 1). Administration of DDC or GF during a longer period did not cause an important accumulation of porphyrins in the bone marrow, while the amount of porphyrins in the plasma remained high. Fluorescing cells were first seen after 3 days treatment with DDC or 7 days with GF. Almost every blood cell showed fluorescence within 2 weeks of administration of these chemicals. Young mice are more sensitive to the porphyrinogenic action of DDC and GF than adult animals.

In quail fluorescing cells were not observed after administration of DDC or GF either in the bone marrow or in the blood. For these experiments very young

TABLE 1

ESTIMATION OF THE AMOUNT OF PP ACCUMULATING IN THE BONE MARROW AND THE BLOOD OF MICE AFTER ADMINISTRATION OF 0.5 % DDC OR 2.5 % GF

		Fluorescence intensity of the PP-spot							
		Cell Content		Plasma		Membranes		Bone marrow	
Hrs	Animal	DDC	GF	DDC	GF	DDC	GF	DDC	GF
24	1	+	++	++++	+++	N.D.	++	N.D.	N.D.
24	2	+	++	+++	+++	N.D.	+	N.D.	N.D.
24	3	+++	+	++++	+++	++	++	N.D.	N.D.
24	4	+++	++	+++	N.D.	+	++	N.D.	N.D.
24	5	+++	+	+++	++	+	+++	N.D.	N.D.
48	6	N.D.	+	+++	+++	N.D.	N.D.	N.D.	N.D.
48	7	++	++	+++	++	+	N.D.	N.D.	N.D.
48	8	+	N.D.	+++	+++	+++	+	N.D.	N.D.
48	9	+++	++	+++	+++	++	+	+++	N.D.
48	10	+++	++	++++	+++	++	+	+	N.D.

N.D.: not detectable
\+ : very weak fluorescence
++ : + < fluorescence intensity < +++
+++ : intensity of the PP-standard (0.1 μg PP)
++++: > standard

quail (5 days after hatching) were used as well as adult animals. In DDC intoxicated quail only the liver and the bile bladder showed an accumulation of porphyrins.

Administration of lead (as lead acetate) caused the occurrence of fluorescing cells in the blood and the bone marrow of Japanese quail within 24 hrs (fig. 1).

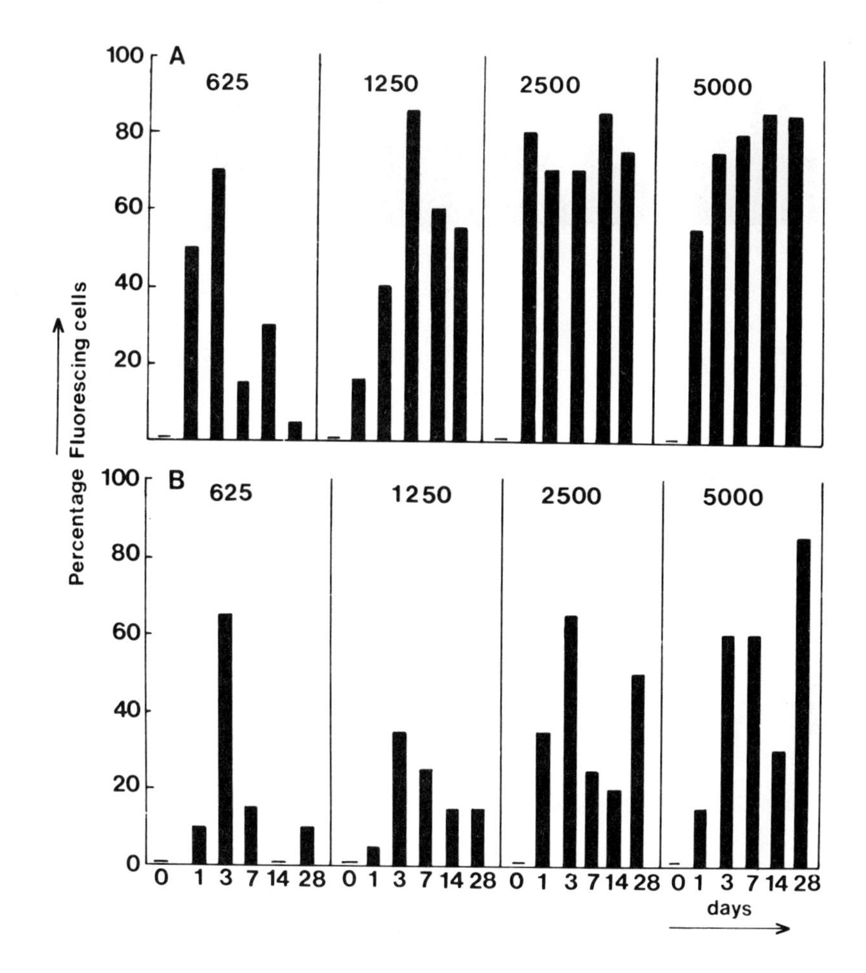

Fig. 1. The percentage of fluorescing cells in the bone marrow (A) and the blood (B) of Japanese quail after administration of 625, 1250, 2500 or 5000 ppm lead in food.

The plasma contained hardly detectable amounts of porphyrins throughout these experiments (fig. 2).

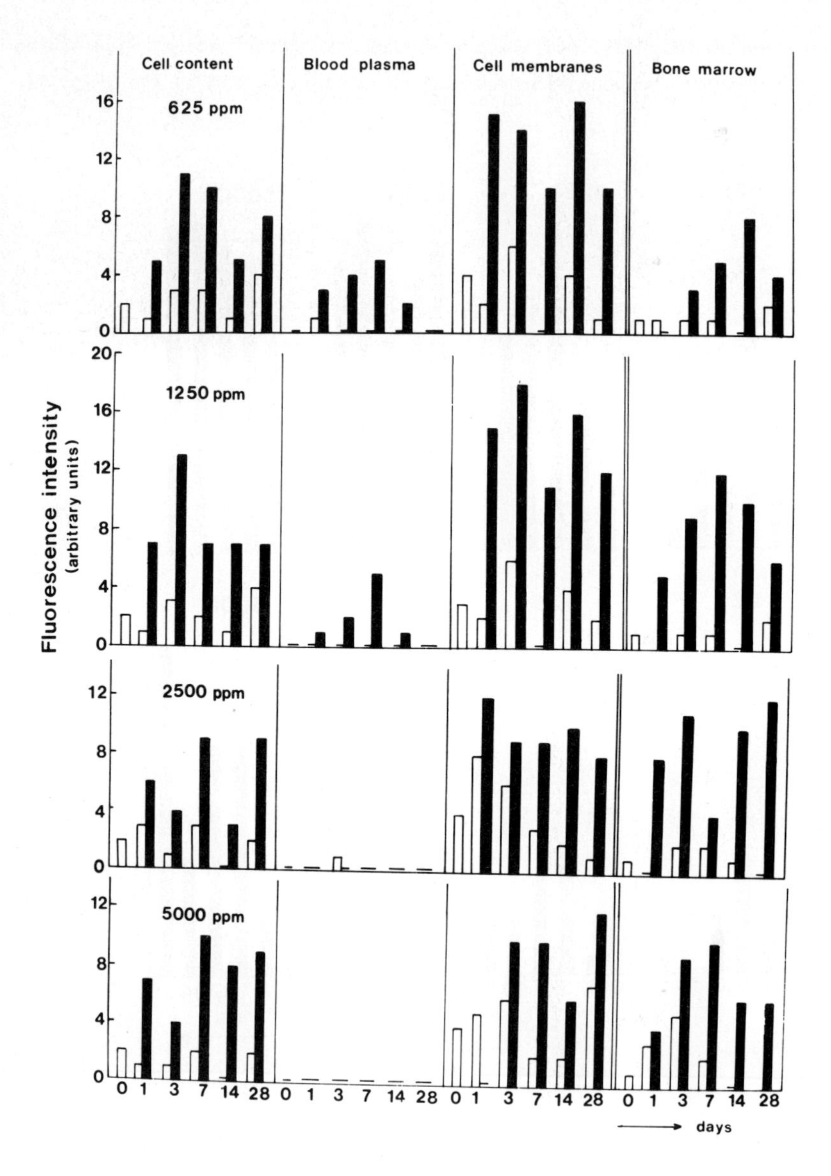

Fig. 2. Estimation of the amount of PP accumulating in the bone marrow and the blood of Japanese quail after administration of lead. Each bar represents the total score of one group (n=5). Empty bars: control. Bark bars: lead-treated animals.

In mice administration of lead caused an elevation of the porphyrin concentration in the spleen (mainly CP). The liver of mice and quail did not show any fluorescence after lead administration. In blood and bone marrow of mice (adult) fluorescing cells were not detectable. We observed some fluorescing cells in the bone marrow of mice after lead administration when we used very young animals (5 weeks), however, during prolonged administration of lead these fluorescing cells disappeared.

Quantitative PP determinations in the blood and bone marrow of DDC treated mice (fig. 3) showed an accumulation of PP in the blood cells which is comparable with the PP concentration in the blood cells of EPP-patients. The PP concentration in the plasma of these animals is more elevated than in EPP-patients.

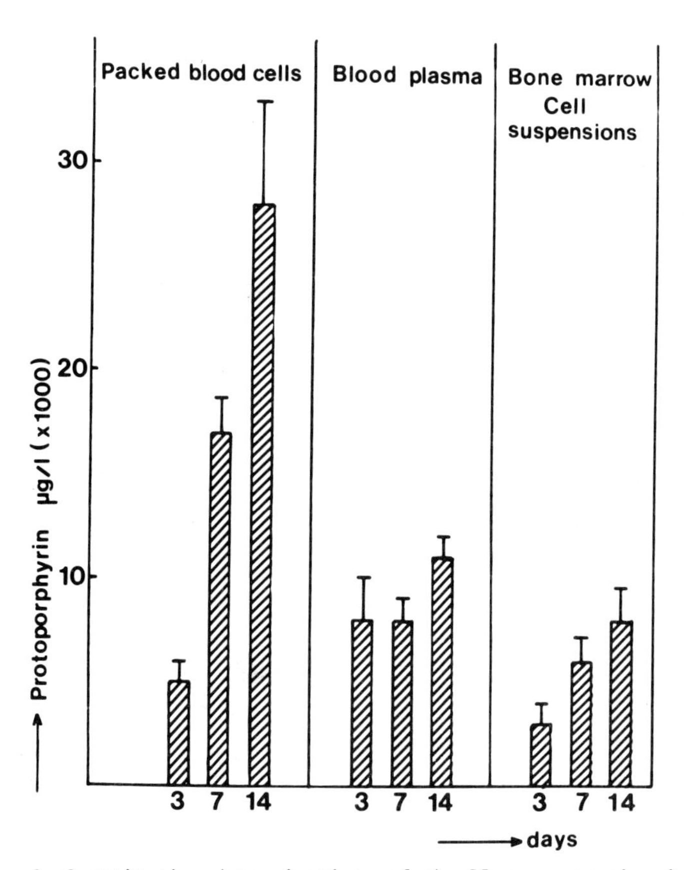

Fig. 3. Quantitative determinations of the PP concentration in the blood cells, the blood plasma and the bone marrow of mice after administration of 0.5 % DDC. Each bar represents the mean of 5 determinations ± S.E.
Empty bars: control. Dark bars: lead-treated animals.

The accumulation of PP in the blood cells and bone marrow cells of lead intoxicated quail was less pronounced than in DDC treated mice (fig. 4).

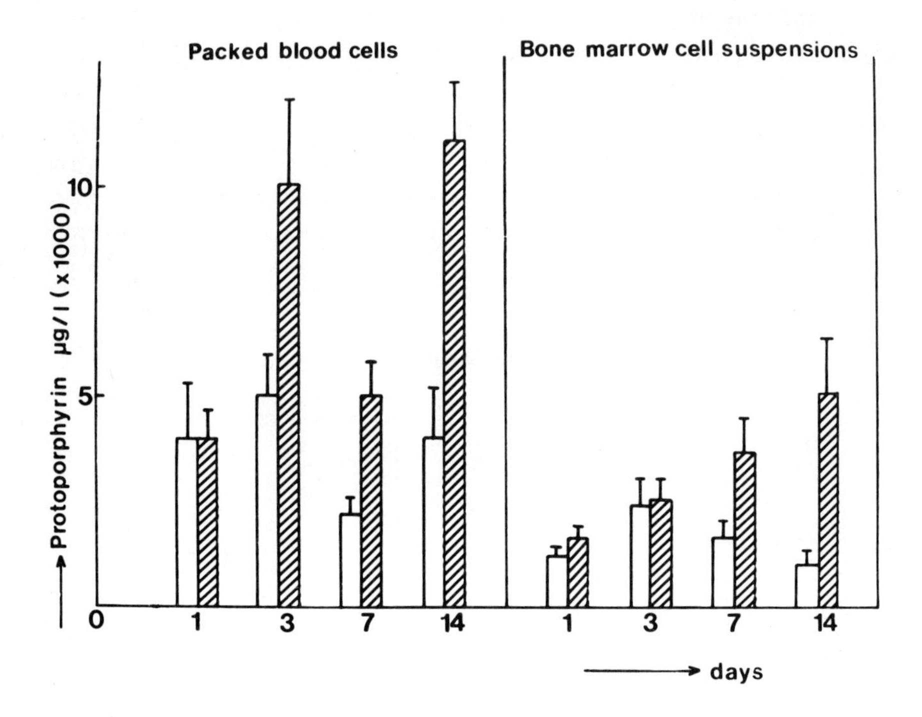

Fig. 4. Quantitative determinations of the PP concentration in the blood cells and the bone marrow of Japanese quail after administration of dietary lead (2500 ppm). Each bar represents the mean of 5 determinations ± S.E. Empty bars: control. Dark bars: lead treated animals.

Photohaemolysis experiments

Table 2 shows that the photohaemolysis of erythrocytes obtained from DDC or GF treated mice is dependent on: 1. the duration of drug administration; 2. the age of the animals; 3. the presence of plasma.

TABLE 2

IRRADIATION INDUCED HAEMOLYSIS OF BLOOD CELLS OBTAINED FROM YOUNG (5-6 WEEKS) AND ADULT (8-12 WEEKS) MICE, AFTER ADMINISTRATION OF 0.5 % DDC OR 2.5 % GF.

0.5 % DDC (5-6 weeks)		0.5 % DDC (8-12 weeks)			2½ % GF (5-6 weeks)		
days	50% haemolysis after (hrs)	days	50% haemolysis after (hrs) - plasma	50% haemolysis after (hrs) + plasma	days	50% haemolysis after (hrs) - plasma	50% haemolysis after (hrs) + plasma
3	7	4	-	-	4	-	ND
5	4½	5	±	-	7	7	ND
7	3	6	7½	-	12	4	6½
14	3	14	4	7	18	4	6½

ND = not done
- = no haemolysis after 8 hrs of irradiation
± = haemolysis less than 50 % after 8 hrs.

The presence of plasma in the suspension always inhibited the effect of haemolysis, in spite of the presence of significant amount of PP in the plasma added! The blood of mice treated with DDC or GF was more sensitive to 410 nm light irradiation than the blood of EPP-patients. The blood of EPP-patients only showed photohaemolysis during irradiation when the concentration of PP was 60,000 μg/l packed cells or more. The blood of lead intoxicated quail (24 hrs) also showed photohaemolysis during irradiation with 410 nm light (within 8 hrs). However, this sensitivity for irradiation decreased with prolonged administration of lead.

Preliminary in vitro experiments

The blood cells and the bone marrow cells of Japanese quail were used to determine porphyrin synthesis in vitro. These blood cells and bone marrow cells synthesize detectable amounts of porphyrins from ALA. Most of the porphyrins accumulated in the medium not in the cells. TLC observations showed a synthesis of mainly PP and UP. CP and other haem precursors were hardly detectable. After 24 and 48 hrs of incubation the suspensions contained fluorescing cells and after 48 hrs the medium was also fluorescing. The bone marrow cell cultures contained many strong fluorescing cells. The blood cells showed a weak fluorescence. Some mature erythrocytes in the blood cell cultures had a strongly fluorescing nucleus. Cell cultures containing bacteria (which were also fluorescing) showed an accumulation in the medium of almost all the haem-precursors (mainly CP).

At this moment we are not able to find significant effects of DDC, GF, lead or other chemicals in this system.

DISCUSSION

The results of our experiments with DDC, GF and lead indicate a different origin for the fluorescing cells in the blood and the bone marrow of animals treated with these chemicals. DDC and GF act primarily porphyrinogenic in the liver. Obstruction of the PP excretion via the bile (PP crystals were observed in the liver within 48 hrs of DDC or GF administration to mice) causes an excretion of PP in the plasma. Accumulation of this PP in circulating blood cells causes the appearance of fluorescing cells in the blood. Since every blood cell may collect from the plasma, all the blood cells might become fluorescing. This suggestion explains our observations that nearly all the blood cells become fluorescing after administration of DDC or GF to mice[22].

Although a deposition of pigment has been described in the livers of Japanese quail[21] the bile ducts seem not to be obstructed, resulting in a normal excretion of PP via the bile. As a consequence of this elevated PP excretion the bile bladder of DDC treated quail is strongly enlarged and shows an intense fluorescence. In addition the PP concentration in the plasma is not elevated and an accumulation of PP in the blood cells is not observed.

Interference by DDC and GF with the haemsynthesis in the bone marrow cells seems unlikely but is not excluded. Our experiments with bone marrow cell cultures provide us with more information regarding this aspect.

The fluorescing cells in the blood of lead-poisoned quail originate from an interference by lead with the haemsynthesis in circulating immature blood cells and bone marrow cells[23]. The observed effect of lead in the spleen of mice causing an accumulation of porphyrins has never described and needs further attention. The absence of fluorescing cells in the blood and the bone marrow of adult mice after administration of lead indicates that the occurrence of fluorescing cells in the blood of animals cannot serve as a parameter for lead poisoning.

Our results are in contrast with the observations of Poh-Fitzpatrick[13]. He observed many fluorescing cells in the blood and the bone marrow of mice after administration of lead (in drinking water) during 1-2 weeks. The discrepancies between our results and those obtained by others may be due to one or more of the following factors: the age of the experimental animals, protein intake, lead dosage and the route of the administration.

Other chemicals like hexachlorobenzene and hexabromobiphenyl induce a hepatic porphyria in quail[21]. Simon[19] described the appearance of plasma fluorescence in HCB-treated rats. In spite of the accumulation of UP in the plasma, fluorescing cells were not observed either in the blood or in the bone marrow of HCB or HBB treated quail (unpublished observation). This absence of fluorescing cells might be due to the inability of UP (and CP?) to accumulate in circulating blood cells. Further experiments in vitro will provide us with more information about these aspects.

The difference between the fluorescing cells in the blood of DDC/GF treated mice, lead treated quail and EPP-patients in their sensitivity for irradiation is probably due to the number of fluorescing cells (the PP concentration seems less important) and the binding of PP to cell membranes or cell proteins. The blood of DDC or GF treated mice contained almost 100 % fluorescing cells. The blood of EPP-patients contained 25 - 40 % fluorescing cells. In the erythrocytes of EPP-patients almost 90 % of the PP seems to be bound to haemoglobin. Our results with DDC and GF indicate that much of the accumulated PP is bound to the cell membranes. These differences in localisation of PP in the cell may play an important role in sensitivity to 410 nm light irradiation.

ACKNOWLEDGEMENTS

Thanks are due to Miss C.Z. Troost and Mr. J.M. Eyndhoven for their assistance in preparing the manuscript.

REFERENCES

1. Allen, B.R., Hunter, J.A.A., Beattie, A.D. and Moore, M.R. (1974) Scot. Med. J., 19, 3.
2. Behrend, G. (1899) Berl. Dtsch. m. Wschr., 25, 254.
3. Cripps, D.J. and MacEachern, W.N. (1971) Arch. Pathol., 91, 497.
4. Doss, M. (1967) J. Chromatogr., 30, 265.
5. Doss, M. (1974) Porphyrins and porphyrin precursors, in: M.Ch. Curtius and M. Roth (eds.) Clinical Biochemistry, Principles and Methods, Walter de Gruyter, Berlin-New York, Vol. II, p. 1323.

6. Fiedler, H. von (1972) Monatsschr., 157, 79.

7. Harber, L.C. (1965) Med. Clin. North Am., 49, 581.

8. Hönigsmann, H., Gschnait, F., Konrad, K., Stingl, G. and Wolff, K. (1976) J. Invest. Dermatol., 66, 188.

9. Konrad, K., Hönigsmann, H., Gschnait, F. and Wolff, K. (1975) J. Invest. Dermatol., 65, 300.

10. deMatteis, F. and Rimington, C. (1963) Br. J. Dermatol., 75, 91.

11. Piomelli, S. (1973) J. Lab. Clin. Med., 81, 932.

12. Piomelli, S., Lamola, A.A., Poh-Fitzpatrick, M.B., Seaman, C. and Harber, L.C. (1975) J. Clin. Invest., 56, 1519.

13. Poh-Fitzpatrick, M.B. and Lamola, A.A. (1977) J. Clin. Invest., 60, 380.

14. Roscoe, D.E., Nielsen, S.W. and Czikowsky, J.A. (1975) Am. J. Vet. Res., 36, 1225.

15. Salomon, H.M. and Figge, F.J.H. (1959) Proc. Soc. Exp. Biol. Med., 100, 583.

16. Sardasai, V.M., Waldmanm J. and Orten, J.M. (1964) Blood, 24, 178.

17. Schmid, R., Hanson, B. and Schwartz, S. (1952) Proc. Soc. Exp. Biol. Med., 79, 459.

18. Schothorst, A.A. (1972) Photohaemolysis caused by protoporphyrin, thesis, Leiden.

19. Simon, N., Dobozy, A. and Berko, Gy. Arch. klin. exp. Dermatol., 238, 38.

20. Stokvis, B.J. (1895) Z. klin. Med., 28, 1.

21. Strik, J.J.T.W.A. (1973) Experimentele leverporfyrie bij vogels, thesis, Utrecht.

22. Woutersen, R.A., Holsteijn, C.W.M. van, and Wit, J.G. Toxic porphyria, I. The origin of fluorescing blood cells in mice after feeding DDC or griseofulvin, submitted for publication.

23. Woutersen, R.A., Holsteijn, C.W.M. van, and Wit, J.G. Toxic porphyria, II. The origin of fluorescing blood cells in Japanese quail and mice after feeding lead acetate, submitted for publication.

Chemical Porphyria in Man, J.J.T.W.A. Strik and J.H. Koeman eds.

TOXICITY OF HEXACHLOROBENZENE WITH SPECIAL REFERENCE TO HEPATIC GLUTATHIONE LEVELS, LIVER NECROSIS, HEPATIC PORPHYRIA AND METABOLITES OF HEXACHLOROBENZENE IN FEMALE RATS FED HEXACHLOROBENZENE AND TREATED WITH PHENOBARBITAL AND DIETHYLMALEATE

P.R.M. KERKLAAN, J.J.T.W.A. STRIK and J.H. KOEMAN
Department of Toxicology, Agricultural University, Wageningen, The Netherlands

SUMMARY

Female rats were fed HCB in combination with a chronic dosage of diethylmaleate and phenobarbital. The results on liver histology, porphyrin excretion and concentration of metabolites of HCB in liver tissue and excreta suggest a protective role of glutathione during HCB metabolism, when a dechlorinated toxic intermediate is formed. This latter compound is believed to be the porphyrinogenic agent.

INTRODUCTION

Hexachlorobenzene, a fungicide and by-product in many industrial processes, has become one of the major environmental pollutants because of its persistance against biodegradation[10]. HCB is known to induce porphyria in birds[19], mammals[13] and humans[1] which is characterized by overproduction of porphyrins with seven, six and five carboxylic groups[15]. Female animals are more prone to develop porphyria[16,9].

All the details of the detoxifying mechanism of HCB in the liver are not yet clear. Evidence has been presented for the protective role of glutathione, during metabolism of lower chlorine-substituted benzenes[6]; an intermediate arene-oxide is conjugated with glutathione and excreted as mercapturate in the urine. Sufficient depletion of glutathione will cause the intermediate to react with tissue macromolecules thus causing centrilobular necrosis in the liver[14]. When co-administered with phenobarbital, a P-450 activating drug, or diethylmaleate, a glutathione depleting agent, the toxic action of these lower chlorinated benzenes is increased[14]. In HCB-treated animals porphyrin fluorescence is accompanied by centrilobular necrosis in the liver[17]. HCB also acts on the drug-metabolizing enzymes while hepatic porphyrin levels are increased[4,16]. Apparently metabolism of HCB is a perequisite for its porphyrinogenic action. Indeed, metabolites of HCB have been detected: pentachlorophenol[11], tetrachlorohydroquinone and pentachlorothiophenol[7].

This study was undertaken to investigate the effects of an induced P-450 system on the metabolism of HCB and the onset of porphyria by administrating HCB simultaneously with phenobarbital to female rats. In a simultaneous experiment, diethylmaleate was given to rats treated with HCB in order to elucidate the role of glutathione.

MATERIALS AND METHODS

Materials

Hexachlorobenzene (analytical grade) was purchased from BDH Ltd., Poole, England. Diethylmaleate was obtained from Merck AG, Darmstadt, West-Germany and phenobarbital from Interpharm BV, Holland.

Animal treatment

Groups of adult female Wistar rats (5 per group) had access *ad libitum* to food containing 0.1% HCB; one-third was fed 1.2% diethylmaleate in the food, another one-third was given 0.1% phenobarbital in drinking-water. Assuming an average daily intake of 15 g food and an average bodyweight of 200 g, this is equivalent to 750 mg HCB and 0.6 ml diethylmaleate per kg bodyweight. Food without HCB was given to control animals, one-third receiving diethylmaleate, and another phenobarbital, respectively. Urine and feces were collected in the second, fourth, sixth and eight' week. After two, four and eight weeks of treatment equal numbers of animals were decapitated, the blood was collected, and the liver was taken out, weighed and divided for histological analysis, estimation of glutathione levels, porphyrin fluorescence and HCB and metabolites analysis.

Analytical procedures

Glutamate-dehydrogenase (GLDH) and glutamate-pyruvate-transaminase (GPT) activities in the serum were measured according to the Boehringer Test Combination. Total porphyrins were determined in urine according to Doss[2]. Liver was scanned microscopically for porphyrin fluorescence under Wood's light (400 nm). A fluorimetric method was used for the estimation of reduced glutathione (GSH) in the liver[5]. Paraffin liver sections were stained with haematoxylin-eosin for histological analysis.

Hexachlorobenzene, pentachlorophenol (PCP), tetrachlorohydroquinone (TCH) and pentachlorothiophenol (PCThP) residues in livertissue, urine and feces were extracted into ether and analysed by gas-chromatography[7]. PCP, TCH and PCThP were methylated prior to analysis. The complete procedure was carried out in the Institute of Toxicology and Pharmacology, Philips University, Marburg, West-Germany under guidance of Dr. G. Koss.

RESULTS

In the course of the experiment the animals in the HCB/diethylmaleate group showed less appetite and decreasing body weight. Consequently, considerations concerning this group with respect to HCB intake should be treated with caution. The remaining groups showed no significant decrease in food consumption. Rats fed HCB/phenobarbital showed considerable loss of hair. Liverweight increase could be observed in all HCB treated groups, as well as in the phenobarbital control group, as is shown by liverweight/bodyweight ratios (Table 1).

TABLE 1

LIVER WEIGHT/BODY WEIGHT RATIOS OF FEMALE RATS TREATED AS DESCRIBED IN MATERIALS AND METHODS;

Mean values obtained from five animals; DM = Diethylmaleate, PB = Phenobarbital.

Group	Exposure time (weeks)		
	2	4	8
HCB	0.039	0.045	0.050
HCB + DM	0.042	0.048	0.061
HCB + PB	0.049	0.049	0.057
control	0.032	0.035	0.029
control DM	0.035	0.033	0.028
control PB	0.040	0.049	0.038

As can be seen in Table 2 serum activities of GLDH and GPT were not increased after eight weeks of treatment.

In Fig. 1 urinary excretion of total porphyrins is shown as a function of the exposure time. It is clearly seen that both phenobarbital and diethylmaleate can enhance the porphyrinogenic action of HCB; within the controls no difference could be observed.

After four weeks of treatment, no microscopic fluorescence of porphyrins in the liver of any animal could be detected.

After eight weeks, however, all animals treated with HCB showed significant fluorescence in their livers, in colour and intensity ranging from bright yellowish-red to dim red. No differentiation could be made between the three different treatments.

Concerning hepatic glutathione concentrations, a slight decrease was observed in rats treated with HCB for two and four weeks, compared to control animals; however, after eight weeks values had returned to control levels as is shown in fig. 2.

After four and eight weeks of treatment, liver sections were examined histologically. Within four weeks, livers of all HCB treated animals showed hypertrophy of the centrilobular cells; no clear difference was seen between the three treatment groups. Aninals treated with HCB for eight weeks showed a more pronounced hypertrophy; when HCB was given along with diethylmaleate this phenomenon was generally observed in the whole liver, with exception of the periportal regions of the liverlobules. Foci of necrotic cells in the centrilobular and midzonal areas were seen. Some less necrotic foci were observed in a number of the lobules in the livers of the HCB/phenobarbital treated animals, along with hypertrophy of the centrilobular and midzonal hepatocytes.

TABLE 2

GLUTAMATE-DEHYDROGENASE, GLUTAMATE-PYRUVATE-TRANSAMINASE ACTIVITY IN SERUM OF FEMALE RATS TREATED AS DESCRIBED IN MATERIALS AND METHODS.

Figures represent mean values of five animals. The range is given between parentheses.

SGLDH activity (U/1)			
Group	Exposure	time	(weeks)
	2	4	8
HCB	3.9(2.8-5.1)	5.8(2.6-8.0)	2.2(1.8-3.3)
HCB + DM	2.6(2.0-5.1)	2.6(1.0-4.0)	4.4(2.7-7.3)
HCB + PB	6.4(3.2-6.7)	6.4(2.0-8.5)	5.5(3.7-8.4)
control	3.0(2.0-4.3)	5.0(1.0-9.4)	7.3(4.7-9.2)
control DM	4.4(3.2-6.7)	3.3(1.0-6.6)	2.7(1.8-3.7)
control PB	3.8(3.5-3.9)	3.6(2.0-6.0)	5.7(2.7-5.7)

SGLDH activity (U/1)			
Group	Exposure	time	(weeks)
	2	4	8
HCB	16.8(13-17)	18.2(12-20)	15.8(13-18)
HCB + DM	16.6(13-18)	15.6(12-20)	16.6(15-20)
HCB + PB	16.4(13-18)	12.6(9-16)	21.8(14-27)
control	17.4(15-20)	16.0(12-20)	18.0(15-21)
control DM	16.6(13-21)	14.8(11-17)	19.8(16-22)
control PB	16.2(13-18)	12.4(10-16)	17.4(12-23)

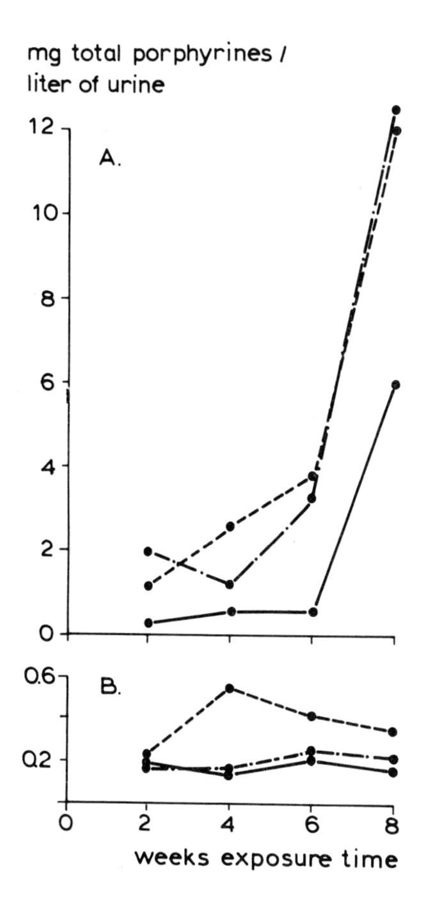

Fig. 1. Urinary total porphyrin excretion; female rats were treated as described under materials and methods. Average values of 2 - 3 animals are depicted. Coproporphyrin was used as an internal standard.
A. HCB treated groups: —— HCB, --- HCB + diethylmaleate, -.- HCB + phenobarbital.

B. control groups: —— control, --- diethylmaleate control, -.- phenobarbital control.

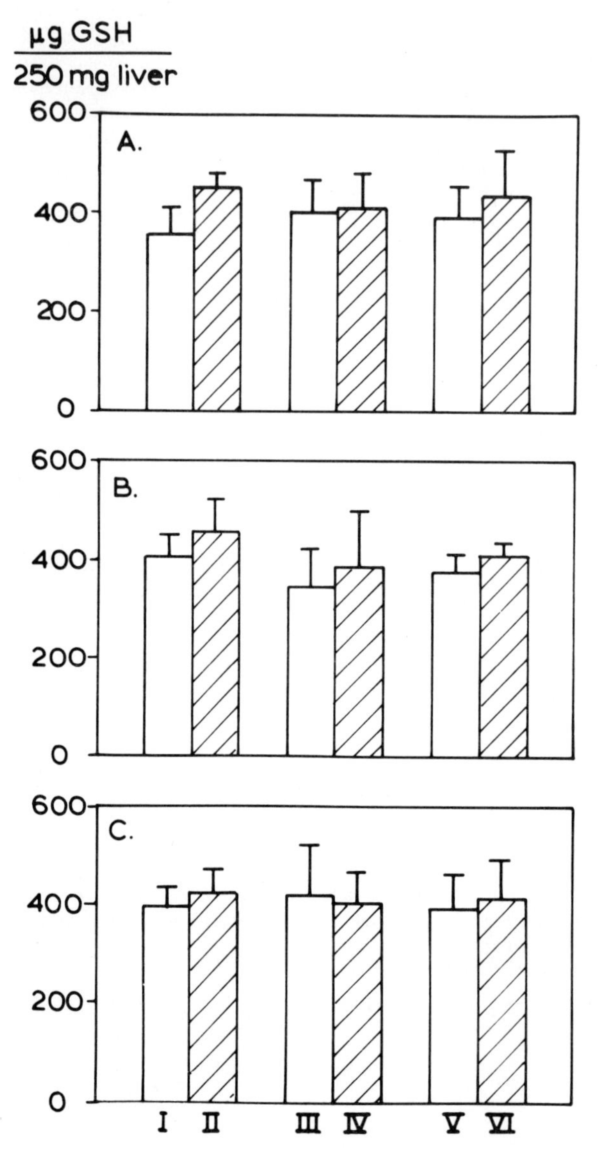

Fig. 2. Hepatic glutathione (GSH) levels of female rats treated as described under materials and methods; I = HCB, II = control, III = HCB + diethylmaleate, IV = diethylmaleate control, V = HCB + phenobarbital, VI = phenobarbital control. A. After two weeks of treatment; B. After four weeks; C. After eight weeks. Figures represents mean values ± S.D. of five animals.

Ethereal extracts of liverhomogenates, urine and feces obtained from animals treated with HCB for eight weeks showed detectable amounts of PCP, TCH and/or PCThP; in urine the concentration of these metabolites was relatively high, while in faeces no TCH and in liver no PCThP could be detected in measurable amounts. As is shown in table 3, the excretion of the sulphur-containing metabolite (PCThP) by animals treated with HCB/diethylmaleate is remarkably low: a ten-fold decrease as compared to the HCB and HCB/phenobarbital treated animals. It is accompanied by a higher excretion of PCP in the urine and unmetabolized HCB in the faeces. No other metabolites were present in detectable amounts.

TABLE 3

RELATIVE CONCENTRATIONS IN PERCENTAGES OF TOTAL RESIDUE LEVEL IN LIVER TISSUE, URINE AND FAECES OF FEMALE RATS TREATED FOR EIGHT WEEKS AS DESCRIBED UNDER MATERIALS AND METHODS.

HCB, PCP, PCThP and TCH were analysed as described under analytical procedures.

	Group	HCB %	PCP %	TCH %	PCThP %	Total residue μMol/g /mg
liver	HCB	98.8	1.01	0.11	--	2.284
	HCB + DM	99.0	0.97	0.01	--	2.768
	HCB + PB	98.7	1.03	0.06	--	1.534
	control	100	--	--	--	0.060
urine	HCB	3.21	56.1	16.7	23.8	0.285
	HCB + DM	0.81	81.0	15.9	2.35	1.233
	HCB + PB	5.05	62.7	19.7	12.4	0.218
	control	100	--	--	--	0.020
faeces	HCB	96.7	2.29	--	1.03	2.524
	HCB + DM	98.6	1.22	--	0.18	5.464
	HCB + PB	91.3	7.53	--	1.13	0.970
	control	100	--	--	--	0.121

DISCUSSION

In the past not much attention has been paid to the role of glutathione in the in vivo metabolism of HCB in relation to chemical porphyria. Wolf et al.[20] reported no significant decrease in hepatic GSH levels in HCB treated, male rats.

Results presented in this paper show that this also applies to female rats.

Phenobarbital and diethylmaleate are able to enhance the porphyrinogenic action of HCB, as is reflected in the marked increase of urinary porphyrin excretion, compared to the action of HCB alone. Since phenobarbital is known to induce cytochrome P-450 and diethylmaleate is known as an glutathione depleting agent, both seem to influence the metabolic fate of HCB in a different way. Since HCB also induces P-450[4,16], the role of phenobarbital is easily understood. By inducing P-450 it adds to the stimulatory effect on HCB metabolism already accomplished by HCB itself, thus promoting extra formation of toxic intermediates and/or metabolites. Concerning the excretion of metabolites, the action of phenobarbital is not reflected in an excretion of larger amounts of metabolites. So far no explanation has been found for this phenomenon. Combined administration of HCB and diethylmaleate lowers the relative amount of PCThP excreted. This sulphur-containing metabolite must be the product of the hydrolysis of dechlorinated hexachlorobenzene mercapturate and so a direct relationship between GSH depletion and the blocking of its conjugation with an electrophilic metabolic product of HCB is evident. The enhanced action of diethylmaleate on HCB-porphyria can now be explained as follows.

A metabolic product of HCB, presumably a strongly reactive intermediate, is responsible for disturbing heme synthesis.

A decreased UROgendecarboxylase activity as a result of HCB treatment has been reported[3], and an interaction between a metabolic product of HCB and the enzyme has been suggested as the cause of porphyria.

Since no large scale necrotic damage of liver cells could be demonstrated (SGPT and SGLDH) levels were not increased) the action of the metabolic product on decarboxylase is presumed to be a direct one. None of the known metabolites of HCB induces porphyria[8] and thus one should conclude that some other metabolic derivative of HCB is responsible. Since the formation of arene-oxides does not seem very likely, enzymatic epoxidation as an intermediate step in HCB metabolism is not likely either. A direct dechlorination of HCB is more probable, as suggested by Mehendale et al.[12], who discovered a directly dechlorinating enzyme in the microsomal fraction of rat livers. Reductive dechlorination as an intermediate step in the metabolism of chlorinated-aromatic hydrocarbons was also reported by Tulp et al.[18].

Finally, since no decrease of hepatic glutathione levels was found during HCB treatment, it is concluded that during long-term exposure the availability of GSH for conjugation is not reflected by its concentration in the liver.

ACKNOWLEDGEMENTS

The authors are very grateful for the technical assistance provided by the Technological Department, the drawnings by C. Rijpma and M. Schimmel and the photography by A. van Baaren of the Biotechnion of the Agricultural University at Wageningen. Typing of the manuscript by Miss G. van Steenbergen and Mrs. L. Muller Kobold - de Lagh is greatly appreciated. We are indebted to the laboratory assistance of Miss E.G.M. Harmsen. The Centre for Small Experimental Animals took care for the housing of the experimental animals. We are indebted to Dr. G. Koss of the Institute of Toxicology and Pharmacology of the Phillips University, Marburg, West Germany where the residue analyses were carried out.

REFERENCES

1. Cam, C. and Nigogosyan, G. (1963) J. Amer. Med. Ass., 183, 88.
2. Doss, M. (1974) Clinical Biochemistry, Principles and Methods, eds. H.Ch. Curtius and M. Roth, De Gruyter, Berlin, p. 1324.
3. Elder, G.H., Evans, J.O. and Matlin, S.A. (1976) Clin. Sci. Mol. Med., 51, 71.
4. Grant, D.L., Iverson, F., Hatina, G.V. and Villeneuve, D.C. (1974) Physiol. Biochem., 4, 159.
5. Hissin, P.J. and Hilf, R. (1976) Anal. Biochem., 74, 214.
6. Jollow, D.J., Mitchell, J.R., Zampaglione, N. and Gilette, J.R. (1974) Pharmacology, 11, 151.
7. Koss, G., Koransky, W. and Steinbach, K. (1976) Arch. Toxic., 35, 107.
8. Koss, G., Seubert, S., Seubert, A., Koransky, W. and Ippen, H. (1977) Abstracts Joint Meeting German and Italian Pharmacologists, Venice, October, 1977.
9. Kuiper-Goodman, T., Grant, D.L., Moodie, C.A., Korsrud, G.O. and Munro, I.C. (1977) Toxic. Appl. Pharm., 40, 529.
10. Leoni, V. and D'Arca, S.U. (1976) Sci. Total. Environ., 5, 253.
11. Lui, H. and Sweeney, G.D. (1975) FEBS Letters, 51, 225.
12. Mehendale, H.M., Fields, M. and Matthews, H.B. (1975) J. Agr. Food Chem., 23, 261.
13. Ockner, R.K. and Schmid, R. (1961) Nature, 189, 499.
14. Reid, W.D. and Krishna, G. (1973) Exp. Molec. Pathol. 18, 80.
15. San Martin de Viale, L.C., Viale, A.A., Nacht, S. and Grinstein, M. (1970) Clin. Chem. Act., 28, 13.
16. Strik, J.J.T.W.A. and Koeman, J.H. (1976) Porphyrinogenic action of hexachlorobenzene and octachlorostyrene. In: Porphyrins in Human Diseases, ed. M. Doss, Karger, Basel, p. 418.
17. Sweeney, G.D., Janigan, D., Mayman, D. and Lai, H. (1971) Sth. Afr. J. Lab. Clin. Med., 45, 68.

18. Tulp, M.Th.M., Bruggeman, W.A. and Hutzinger, O. (1977) Separ. Exp. 33, 1135.

19. Vos, J.G., Breeman, H.A. and Benschop, H. (1968) Meded. Rijksfac. Landb. wetensch. Gent, 33, 1236.

20. Wolf, M.A., Lester, R. and Schmid, R. (1962) Biochem. Biophys. Res. Comm., 8, 278.

Chemical Porphyria in Man, J.J.T.W.A. Strik and J.H. Koeman eds.

TOXICITY OF HEXACHLOROBENZENE (HCB) WITH SPECIAL REFERENCE TO HEPATIC P-450 LEVELS, P-450 BINDING AFFINITIES AND D-GLUCARIC ACID, MERCAPTURIC ACID AND PORPHYRIN LEVELS IN URINE OF FEMALE RATS FED HCB AND TREATED BY PHENOBARBITAL (PB) AND DIETHYLMALEATE (DH)

L. PUŻYŃSKA[a], F.M.H. DEBETS[b], J.J.T.W.A. STRIK[b]

[a]Food and Nutrition Institute, Warsaw, Poland.

[b]Department of Toxicology, Agricultural University, Wageningen, The Netherlands

SUMMARY

In female Wistar rats treated with hexachlorobenzene (HCB) given with or without diethylmaleate (DM) or phenobarbital (PB) the following changes were found:

(1) the elevated level of urinary D-glucaric acid in rats receiving HCB together with DM or to a lesser degree in those receiving HCB in conjunction with PB;

(2) more mercapturates in the urine of rats treated with HCB given together with PB and especially with DM;

(3) no disturbances in the excretory kidney function in experimental rats;

(4) the increased concentration of cytochrome P-450 in the microsomal fraction from the liver of rats fed HCB plus DM, HCB alone and HCB plus PB;

(5) the p-nitroanisole O-demethylation reaction was about 2 times greater in the hepatic microsomal fraction of the group given HCB and in the HCB plus DM treated group, when compared to the corresponding controls. In the HCB plus PB treated group no elevation was found;

(6) the K_s-value for aniline was decreased to the same extent when measured with microsomes from rats treated with HCB, HCB plus DM and PB, and to a greater extent when microsomes from rats treated with HCB plus PB were used. The maximum spectral change for aniline, which is a measure of the total number of binding sites, was increased to the same extent in microsomes from rats treated with HCB, HCB plus DM and HCB plus PB.

It seems from these results that DM, by competing for hepatic GSH, and PB, by stimulation of HCB metabolism, exert influence on the toxic action of intermediates formed during HCB metabolism.

INTRODUCTION

It is well known that many foreign compounds of different chemical structure, after penetrating into the body, stimulate increased production of the enzymes which participate in their metabolism. Among these enzymes, a very important group represent those oxidoreductases in the liver endoplasmic reticulum that depend on NADPH and cytochrome P-450. This response of the liver to foreign inducing substances is often parallelled by a stimulation of hepatic carbohydrate metabolism via the glucuronic acid pathway which is reflected by an elevated excretion of L-absorbic acid and D-glucaric acid (D-GA) in the urine. Therefore, the degree of activity of these liver drug-metabolizing enzymes may readily be estimated by determining D-GA in the urine. This was pointed out by Marsh and Reid[1] and fully confirmed by many others[2,3,4]. At present excretion of D-glucaric acid in urine is considered as an indirect sensitive index of enzyme induction and a useful nonspecific indicator, in man also, of an intake of certain compounds foreign to the organism.

The second useful parameter, indicating the exposure of the organism to foreign substances, especially those with potentially alkylating properties which are most hazardous, may be measuring of urinary concentration of mercapturates. Mercapturate formation appears to involve only foreign compounds; no case of a derivative formed from an endogenous substance has been reported. Exceptions are prostaglandins and estrogens at extremely low concentrations[5,6,7]. Many foreign compounds are in part converted to mercapturates and it is one of the routes of their detoxication and a primary protective mechanism of the organism against their harmful effects. Especially dangerous are electrophilic agents which occur as such or can arise as reactive intermediates through the action of e.g. the mixed-function oxidase system. Mercapturate formation occurs by conjugation with glutathione (GSH), both enzymatically and non-enzymatically. Glutathione conjugates are further metabolized by subsequent cleavage of the glutamate and glycine moieties, followed by acetylation of the free amino group of the cysteine residue. The final product of this metabolic pathway - a mercapturate - is subsequently excreted in the urine where its concentration may provide an indication of the presence of very toxic alkylating compounds in the body.

Hexachlorobenzene (HCB) belongs to the group of chronic toxic, not readily metabolized, foreign substances inducing numerous damages in the organism. Because its metabolism and the mechanism of its toxic action are not well known it seemed that it would be interesting to investigate the possible effect of HCB, given alone or in combination with diethylmaleate (DM) or phenobarbital (PB) on:

1) urinary excretion of D-glucaric acid, and concentration of cytochrome P-450 in the microsomal fraction of the liver. For it is obvious now that the concentration of this carbon monoxide-binding hemoprotein in liver microsomes increases also when animals are treated with stimulators of drug-metabolizing enzymes;
2) urinary concentration of mercapturic acid and other thioethers. Koss et al.[8], identified pentachlorothiophenol and tetrachlorothiophenol as HCB metabolites indicating that -SH groups are involved in its metabolism.

Being metabolized by mercapturic acid biosynthesis DM depletes the hepatic GSH levels. So HCB metabolism by mercapturic acid formation should be decreased if GSH is really involved in the metabolism of this compound.

PB, being a known inducer of the drug-metabolizing system, should promote HCB metabolism and cause formation of more metabolites.

MATERIALS AND METHODS

Female Wistar rats were treated with HCB, DM and PB in the way described by Kerklaan[9].

The estimation of D-glucaric acid and mercapturic acid was carried out on pooled samples (2 or 3 animals) from 24-hour urine collection. The urine collection was done after 2, 4, 6 and 8 weeks of experiment. Urine which could not be analyzed on the day of collection was stored at -50 °C.

D-glucaric acid was measured according to Marsh[1] from the inhibitory effect of D-glucaro-1,4-lactone, to which it is converted by heating at pH 2.0, on β-glucuronidase (EC 3.2.1. 31).

Mercapturic acid concentration was estimated according to the method described by Seutter-Berlage et al.[10] with modifications given personally. The free SH-groups and those liberated during alkaline hydrolysis were assayed by the method of Ellman[11] using 5,5'-dithiobis-(2-nitrobenzoic acid).

The urine creatinine concentration was also measured to verify completeness of the urine collection. After 8 weeks of experiment the creatinine level in the serum of the blood was measured as well in order to assess the kidney excretory function. Creatinine in the urine and serum was measured by the method of Jaffè[12], based on the colour reaction between creatinine and alkaline picrate reagent.

The concentration of D-glucaric acid was expressed in mg of glucaric acid per mmol of creatinine.

The level of mercapturic acid and other thioether compounds in urine was expressed as a molar ratio SH- to creatinine.

Preparation of microsomes: animals were killed by decapitation, 8 weeks after the start of the experiment. The livers were rapidly excised into ice-cold 0.25M sucrose-50mM Tris-HCl buffer (pH 7.4), scissor-chopped, and rinsed three times with the same buffer. Liver homogenate (1:3, w/v) from the pooled livers of five rats was prepared in 50mM Tris-HCl buffer (pH 7.4) containing 0.25 M sucrose in a Potter-Elvehjem homogenizer equipped with a teflon pestle. The homogenization procedure was carried out in a cold room (0-4 ^{o}C). The 9000g (for 15 min.) supernatant fraction from the centrifuged liver homogenate was recentrifuged at 105,000g for 60 min. The ensuing microsomal pellet was washed by rehomogenization in 0.25M sucrose-50 mM Tris-HCl buffer (pH 7.4) and recentrifuged at 105,000g for 60 min. The final, washed microsomal pellet was resuspended to a concentration equivalent of 0.5 - 1 g of liver (wet weight) per ml in 50mM Tris-HCl buffer (pH 7.4) containing 0.25M sucrose and stored in liquid nitrogen (-196 ^{o}C) until using it for further determinations. In microsomal suspension, done also with Tris-sucrose buffer, (pH 7.4) the level of cytochrome P-450 and b_5(nmole) was determined according to the method of Omura and Sato[13] and expressed per mg of microsomal protein. Protein was estimated by Lowry's method[14] with bovine serum albumin as a standard. Spectral interaction with cytochrome P-450 of the following substrates: aniline, hexachlorobenzene, hexobarbital, phenobarbital, cyclohexane and p-nitroanisol - was determined as well. The difference spectra were obtained by addition of microliter quantities of a solution of the ligand to the sample cuvette. Spectra were recorded in the dual-beam mode of a Beckman model Acta C III UV-Vis spectrophotometer in the manner described by Schenkman et al.[15].

Microsomal O-dealkylation was determined by the demethylation of p-nitroanisole. The formed p-nitrophenol was measured by the method of Netter[16].

RESULTS

Urinary excretion of D-glucaric acid

As can be seen from table 1 HCB or DM given alone did not influence the concentration of D-glucaric acid in urine. On the other hand the amount of D-glucaric acid was increased in urine of rats poisoned by HCB given together with DM. This increase reached about 257 % after 4 and 6 weeks of exposure and 378 % after 8 weeks, when compared with the value obtained after 2 weeks. This concentration of D-glucaric acid was in the same range as found in control rats or rats given only DM or HCB. Similarly, excretion of D-glucaric acid in urine was increased in rats given PB - a known inducer of drug-metabolizing enzymes. This increase amounted to 303 % in comparison with the value obtained in control rats. The increased level of D-glucaric acid, found in the urine of these rats after 2 weeks, remained almost constant for the whole period of the experiment. In rats obtaining HCB together with PB the increase in the concentration of D-glucaric acid, in comparison with the values found in control rats, was only a little greater than in rats treated with PB only, especially after 2 and 4 weeks of experiment. This increase was more pronounced in these rats after 6 (210 %) and 8 (403 %) weeks of experiment. It confirms that HCB alone does not cause the evident increase of D-glucaric acid in urine.

TABLE 1

THE LEVEL OF D-GLUCARIC ACID, EXPRESSED IN mg/mMol OF CREATININE, IN THE URINE OF FEMALE RATS FED HCB AND TREATED WITH PB OR DM;

HCB and DM were given in the standard diet in the concentrations of 1000 ppm and 12 ppm respectively. The dose of DM was raised 100 times after 17 days of the experiment. PB was given in drinking water in 0.1 % concentration. The estimation was done in the urine collected after 2, 4, 6 and 8 weeks of the experiment.

Group of rats	Time of exposure (weeks)			
	2	4	6	8
I HCB	4.1	4.1	4.2	4.7
II HCB + DM	4.7	18.0	15.6	22.5
III HCB + PB	15.8	11.8	18.0	15.1
IV Control	4.2	6.0	5.8	3.0
V DM	3.0	4.3	7.4	2.6
VI PB	12.0	10.3	9.8	12.1

The level of cytochrome P-450 and b_5 and the activity of p-nitroanisole O-demethylase in the microsomal fraction of the liver of experimental rats.

In spite of the fact that HCB did not promote the increased production of D-glucaric acid it increased the level of cytochrome P-450 (by about 60 %) and cytochrome b_5 (by about 25 %), in the microsomal fraction of rats poisoned for 8 weeks with this compound (table 2), when compared with the values found in control, non-treated rats.

In rats fed diet contaminated by HCB together with DM the level of cytochrome P-450 (but not b_5) was still increased to 109 %. However, it is evident that this increase was caused by HCB or its metabolite(s) formed in the presence of DM, because DM given alone had no influence on either cytochrome P-450 or b_5.

The level of cytochrome P-450 in the rats treated with PB was increased to the same extent as in rats treated with HCB in conjunction with DM. But surprisingly, in rats receiving PB together with HCB the level of cytochrome P-450 was increased only by 43 %. The concentration of cytochrome b_5 was also more increased when PB was given alone (by 50 %) than when it was given together with HCB (by 25 %).

The microsomal cytochrome P-450 - mediated O-demethylation of p-nitroanisole was altered in the same way as the cytochrome P-450 content in the differently treated groups of rats (table 2).

TABLE 2

THE LEVEL OF CYTOCHROME P-450, b_5 AND p-NITROANISOLE O-DEMETHYLASE ACTIVITY IN THE MICROSOMAL FRACTION OF THE LIVER OF FEMALE RATS TREATED FOR 8 WEEKS WITH HCB, DM OR PB, GIVEN ALONE OR IN COMBINATION

HCB and DM were given in the standard diet in the concentration of 1000 ppm and 12 ppm respectively. The dose of DM was raised 100 times after 17 days of the experiment. PB was given in drinking water in 0.1 % concentration.

Group of rats	Microsomal cytochrome P-450 (nmol/mg microsomal protein)	Microsomal cytochrome b_5 (nmol/mg microsomal protein)	p-nitroanisole O-demethylation (nmol p-nitrophenol formed/min/mg microsomal protein)
I HCB	1.20	0.73	4.14
II HCB + DM	1.55	0.55	4.61
III HCB + PB	1.06	0.72	3.68
IV Control	0.74	0.58	1.85
V DM	0.75	0.54	1.85
VI PB	1.52	0.88	4.67

Spectral interactions of different substrates with cytochrome P-450

The trials to obtain Type I spectrum were without success regarding the following substrates, (with concentrations used given in parentheses): HCB (1 mM and saturated solution); phenobarbital and hexobarbital (0.1 mM, 1 mM, 5 mM and 10 mM); cyclohexane and p-nitroanisole (10 mM).

The K_s-value for aniline, a compound causing type II difference spectrum, was decreased to the same extent (by about 75 %) in the microsomal fraction obtained from the liver of rats treated with HCB, HCB plus DM or with PB only (table 3). The K_s-value for aniline was even more decreased (by about 85 %) in

TABLE 3

THE SPECTRAL DISSOCIATION CONSTANT (K_s) AND MAXIMUM SPECTRAL CHANGE (ΔA_{max}) OF ANILINE IN THE LIVER MICROSOMES FROM FEMALE RATS TREATED FOR 8 WEEKS WITH HCB, DM OR PB GIVEN ALONE OR IN COMBINATION

HCB and DM were given in the standard diet in the concentration of 1000 ppm and 12 ppm respectively. The dose of DM was raised 100 times after 17 days of the experiment. PB was given in drinking water in 0.1 % concentration.

Group of rats	Spectral dissociation constant (K_s; in mM)	Maximum spectral change ($\Delta A_{max}/\Delta A_{390-430}$/ nmol P-450)
I HCB	0.11	0.047
II HCB + DM	0.11	0.054
III HCB + PB	0.06	0.042
IV Control	0.43	0.028
V DM	0.67	0.019
VI PB	0.13	0.016

the microsomal liver fraction from the rats given HCB together with PB.

The maximum spectral change (ΔA_{max}) for aniline was increased to the same extent (by about 70 %) in all groups treated with HCB in comparison with the results from microsomes obtained from the livers of control non-treated rats. In contrast with this, the ΔA_{max} values measured in microsomes from DM and PB treated rats, were decreased (table 3).

Urinary mercapturic acid excretion

Evidently the urinary mercapturate excretion, as it is seen from the table 4, did not change under the influence of HCB throughout the experiment. In comparison with values found in control rats the mercapturate excretion in rats treated with HCB seemed to be higher, by 100 %, after only 8 weeks. In rats given DM the level of mercapturates in the urine was already increased by 144 %, after only 4 weeks, after 6 weeks by 137 % and 8 weeks by 475 %, above the values found in the urine of control rats. When HCB was given together with DM, more mercapturates were found in the urine. The concentrations rose to 95 %, 340 %, 840 % and 820 % above the values of control rats, after 2, 4, 6 and 8 weeks respectively.

In the urine of rats treated with PB the concentration of mercapturates was similar to those found in the urine of control rats; but after only 2 weeks of

TABLE 4

URINARY LEVELS OF MERCAPTURIC ACID (AND/OR OTHER THIOETHERS) EXPRESSED AS A MOLAR RATIO SH-(CREATININE) IN FEMALE RATS TREATED WITH HCB, DM OR PB GIVEN ALONE OR IN COMBINATION

HCB and DM were given in the standard diet in the concentration of 1000 ppm and 12 ppm respectively. The dose of DM was raised 100 times after 17 days of experiment. PB was given in drinking water in 0.1 % concentration. The estimation was done with the urine collected after 2, 4, 6 and 8 weeks of the experiment.

Group of rats	Time of exposure (weeks)			
	2	4	6	8
I HCB	0.033	0.045	0.017	0.040
II HCB + DM	0.043	0.220	0.724	0.184
III HCB + PB	0.083	0.067	0.010	0.044
IV Control	0.022	0.050	0.077	0.020
V DM	0.020	0.122	0.183	0.115
VI PB	0.107	0.047	0.079	0.014

PB treatment this value was unexpectedly higher (by 386 %). When HCB was given with PB more thioethers were found after only 2 weeks (by about 277 %) and 8 weeks (100 %) in comparison to control rats. After 6 weeks in this group of rats, as well as in the group obtaining HCB alone, the amount of thioethers found in urine was evidently smaller than in the control group.

Creatinine concentrations in the serum and urine

Eight weeks after the beginning of the experiment, there were no distinct differences between the different groups of rats, when rats of each of the groups were measured with regard to the ratio of serum creatinine level to urine creatinine level. It was equal to 0.028; 0.017; 0.026; 0.014; 0.028 and 0.032 in the following groups respectively: control, HCB treated, DM treated, HCB plus DM treated, PB and HCB plus PB treated. The very small values of these ratios indicate that excretory function of the kidney is not disturbed even after 8 weeks of feeding rats diet contaminated with HCB, either with or without DM or PB.

DISCUSSION

The finding that HCB did not increase the urinary excretion of D-glucaric acid is in contrast with the results obtained by Notten et al.[18]. They showed that HCB given i.p. to guinea pigs at a dose of 10 mg per kg of body weight per day for 21 days caused a significant increase in D-glucaric acid excretion after 7 days of the experiment. Perhaps this discrepancy is due to the different animals used, the dose used, and the way HCB was administered.

Notten et al.[18,19] also demonstrated that stimulation of the D-glucuronate pathway, leading to an enhanced D-glucaric acid excretion, is not necessarily related to an induction of the liver drug-metabolizing enzymes and vice versa. For example, HCB in their experiment did not affect the activity of microsomal hepatic N-demethylase of aminopyrine. Differences between the stimulatory effects of xenobiotic compounds on the drug-metabolizing enzymes and the glucuronic acid pathway have also been reported by Aarts[3].

In our experiment HCB increased the level of cytochrome P-450 and stimulated the microsomal O-dealkylation of p-nitroanisole, but was without any influence on D-glucaric acid excretion. An increased level of cytochrome P-450 and b_5 under the influence of HCB was also demonstrated in the experiments of Wada et al.[20] and Mehendale et al.[21]. The latter authors also found enhanced activity of aniline hydroxylase, ethyl morphine and p-nitroanisole demethylase in the microsomal fraction isolated from the liver of rats treated with HCB.

The increased excretion of D-glucaric acid in the urine of rats receiving HCB together with DM, especially noticeable after 8 weeks, suggests that this stimulation of the D-glucuronic acid pathway may be caused by a metabolite or metabolites of HCB which can not combine with GSH, due to its lowered level caused by DM. Consequently these may accumulate in the liver and stimulate carbohydrate metabolism via the D-glucuronic acid pathway.

A similar situation may exist in rats given HCB together with PB. By inducing HCB metabolism, PB may stimulate the formation of its metabolite or metabolites in amounts for which the liver has not sufficient available GSH. This may also be the reason for their accumulation. This metabolite (or metabolites) may play the main role in causing porphyria in rats, because after 8 weeks of the experiment the amount of porphyrins in the urine was greatest in rats obtaining HCB with PB or DM. But in the liver of these rats the main HCB metabolite was pentachlorophenol and to a lesser degree tetrachlorohydroquinone[9], which are known not to cause any porphyria in rats[22,23] or Japanese quail.

In the urine of rats treated with HCB together with PB and especially together with DM more pentachlorophenol was found than in rats obtaining HCB alone[9]. This metabolite is probably excreted mainly as glucuronide, so its increased formation may also suggest that stimulation of the glucuronic acid pathway had taken place. This is especially so in the light of work done by Mehendale et al.[21] and Notten et al.[18], who found that HCB is a good inducer of UDP-glucuronyltransferase activity, associated with the microsomal fraction. However, the formation of glucuronides may result in less UDP-glucuronic acid being converted to D-glucuronic acid and subsequently to D-glucaric acid. On the other hand it is known that D-glucaric acid has an inhibitory action on the activity of β-glucuronidase, which releases free glucuronic acid from glucuronides. So increased production of D-glucaric acid may cause more glucuronides to be formed and excreted. But, according to the results of Notten et al.[19], stimulation of the D-glucuronate pathway and accelerated glucuronidation are not necessarily related.

HCB increases the level of liver microsomal cytochrome P-450 and b_5. The level of cytochrome P-450 is more increased when HCB is given together with DM, which alone has no influence on the concentration of these hemoproteins. When HCB was given in connection with PB, the level of cytochrome P-450 was increased to a lesser extent in comparison with the value obtained after treatment by PB alone or HCB plus DM. It is tempting to speculate that the synthesis of cytochrome P-450 is influenced not only by HCB, but also by its metabolites. Those metabolites, which are formed in the presence of DM, seem to promote this synthesis, whereas those which are formed in the presence of PB, representing also the substrate of cytochrome P-450, do not.

It is not surprising that our experiment did not succeed in obtaining type I binding spectrum. It is well known that a sex difference in the binding properties of cytochrome P-450 exists in rats. No difference is observed between mature male and female rats in the way in which they oxidize aniline or combine aniline to form the type II spectrum. However, the oxidation of aminopyrine, a type I drug, and the binding of aminopyrine and hexobarbital (type I) is much greater with microsomes from male rats than with microsomes from female rats[4]. Besides, it is also established that microsomes lose some of their ability to metabolize certain drugs when stored in the cold. This is probably due to the loss of the type I binding site. Storage of microsomes at -5 ^{o}C for seven days caused almost complete loss of type I binding, but had little or no effect on type II binding[5]. In the case of this present experiment the microsomes lost their ability to form the type I spectrum in spite of the fact that they were kept in liquid nitrogen. Probably microsomes from female rats are much more sensitive or perhaps the spin state of P-450 iron is changed during storage of the microsomes in liquid nitrogen in such a way that it was impossible to detect a type I difference spectrum. A relationship has been reported between starting in vivo spin state of P-450 iron and changes in difference spectra[7]. A high percentage of high spin P-450 iron gives a good type II difference spectrum but a type I spectrum can hardly be detected. On the other hand there was no loss of cytochrome P-450 and the microsomes kept their capacity to metabolize p-nitroanisole (a type I substrate) after storage in liquid nitrogen.

From the results obtained after measuring the urinary mercapturate content in experimental groups of rats it is evident that the amount of -SH groups from thioethers was elevated in HCB-treated rats after only 8 weeks. Also after this time the percentage of pentachlorothiophenol - sulphur containing metabolite of HCB - in the urine of these rats was the highest in that group in comparison with the urine obtained from rats given HCB plus PB or HCB plus DM[9]. In the case of PB it may be concluded that this substance only promotes the metabolism of HCB to a very small degree, through mercapturate formation. When HCB was given together with DM more mercapturates were formed than when DM was given alone. So it may be concluded that they are probably derived not only from DM metabolism but also from the metabolism of HCB. Surprisingly however, in the urine of the rats of this group the percentage of pentachlorothiophenol was the smallest. So it may also be possible that DM diminished HCB metabolism to SH-containing metabolites.

Finally it may be summarized that the above results suggest that PB, by stimulating HCB metabolism, and DM, by competing for hepatic glutathione, have an influence on the rate of formation of HCB metabolites and the toxic action they exert.

ACKNOWLEDGEMENTS

The authors are very grateful for the technical assistance provided by the Technological Department, the drawnings by C. Rijpma and M. Schimmel and the photography by A. van Baaren of the Biotechnion of the Agricultural University at Wageningen. Typing of the manuscript by Miss G. van Steenbergen and Mrs. L. Muller Kobold - de Lagh is greatly appreciated. We are indebted to the laboratory assistance of Miss E.G.M. Harmsen. The Centre for Small Experimental Animals took care for the housing of the experimental animals.

REFERENCES

1. Marsh, C.A. and Reid, L.M. (1963) Biochem. Biophys. Acta, 78, 726.
2. Aarts, E.M. (1965) Biochem. Pharmacol. 14, 359.

3. Aarts, E.M. (1971) Lancet I, 859.
4. Hunter, J., Maxwell, J.D., Stewart, D.A. and Williams, R. (1973) Biochem. Pharmacol., 22, 743.
5. Wood, J.L. (1970) Biochemistry of mercapturic acid formation, in: "Metabolic conjugation and metabolic hydrolysis", W.H. Fishman (ed.), vol.II, Academic Press, New York, p. 261.
6. Jocelyn, P.C. (1972) Reaction of thiols involving their consumption, in: Biochemistry of the SH-group, Academic Press, London, p. 212.
7. Chasseaud, L.F. (1976) Conjugation with glutathione and mercapturic acid excretion, in Glutathione: metabolism and function, J.M. Arias and W.B. Jakoby (eds.), Raven Press, New York, p. 77.
8. Koss, G., Koransky, W. and Steinbach, K. (1976) Arch. Toxicol., 35, 107.
9. Kerklaan, P.R.M., Strik, J.J.T.W.A. and Koeman, J.H. Toxicity of hexachlorobenzene with special reference to hepatic glutathione levels, liver necrosis, hepatic porphyria and metabolites of hexachlorobenzene in female rats fed hexachlorobenzene and treated with phenobarbital and diethylmaleate. This volume.
10. Seutter-Berlage, F., Dorp, H.L. van, Kosse, H.G.J. and Henderson, P.Th. (1977) Int. Arch. Occup. Environ. Hlth., 39, 45.
11. Ellman, G.L. (1959) Arch. Biochem. Biophys., 82, 70.
12. Jaffé, M. (1886) Z. Physiol. Chem., 10, 391.
13. Omura, T. and Sato, R. (1964) J. Biol. Chem. 239, 2370.
14. Lowry, O.H., Rosenbrough, N.J., Farr, A.L. and Randall, R.J. (1951) J. Biol. Chem., 193, 265.
15. Schenkman, J.B., Frey, J., Remmer, H. and Estabrook, R.W. (1967) Mol. Pharmacol., 3, 516.
16. Netter, K.J. (1960) Schmiedebergs Arch. Exp. Path. Pharmak., 238, 292.
17. Nebert, D.W., Kumaki, K., Sato, M. and Kon, H. (1977) Association of type I, type II and reverse type I difference spectra with absolute spin state of cytochrome P-450 iron, in: "Microsomes and drug oxidations", Proc. of the third Int. Symposium, Berlin, V. Ullrich (ed.), Pergamon Press, Oxford, 224.
18. Notten, W.R.F. and Henderson, P.Th. (1975) Alterations in urinary D-glucaric acid excretion as an indication of exposition to xenobiotics, in: "Alterations in the D-glucuronic acid pathway and drug metabolism by exogenous compounds", W.R.F. Notten, doctoral thesis, Nijmegen.
19. Notten, W.R.F. and Henderson, P.Th. (1975) Int. J. Biochem., 6, 111.
20. Wada, O., Yano, Y., Urata, G. and Nakao, K. (1968) Biochem. Pharmacol., 17, 595.
21. Mehendale, H.M., Fields, M., Matthews, H.B. (1975) J. Agr. Food Chem., 23, 261.
22. Goldstein, J.A., Friesen, M., Linder, R.E., Hickman, P., Hass, J.R., Bergman, H. (1977) Biochem. Pharmacol., 26, 1549.
23. Koss, G., Seubert, S., Seubert, A., Koransky, W. and Ippen, H. (1977) On the induction of porphyria by metabolites of hexachlorobenzene, in: Abstracts of the Joint Meeting of German and Italian Pharmacologists, p. 129.

24. Schenkman, J.B., Remmer, H. and Estabrook, R.W. (1967) Mol. Pharmacol. 3, 113.

25. Shoeman, D.W., Chaplin, M.D. and Mannering, G.J. (1969) Mol. Pharmacol., 5, 412.

Chemical Porphyria in Man, J.J.T.W.A. Strik and J.H. Koeman eds.

ON THE EFFECTS OF THE METABOLITES OF HEXACHLOROBENZENE

G. KOSS[a], S. SEUBERT[b], A. SEUBERT[b], W. KORANSKY[a], H. IPPEN[b] and M. STRAUB[b]

[a]Institute of Toxicology and Pharmacology, Philipps-University, Marburg, Fed. Rep. Germany

[b]Clinic of Dermatology, Georg-August-University, Göttingen, Fed. Rep. Germany

G.K. is supported by the Deutsche Forschungsgemeinschaft
A.S. is supported by the P.-G.-Unna Stiftung

SUMMARY

In order to characterize the role of the metabolites of hexachlorobenzene in the induction of toxic effects, like porphyria, pentachlorophenol, pentachlorothiophenol, pentachlorothioanisol and its sulfoxide and sulfone were administered to female rats for several weeks. The gain in body weight and the weight of the liver, spleen and kidney were found not to be influenced by these substances when compared with controls. Moreover, the substances tested neither altered the content of the liver and urinary porphyrins, the content of the urinary delta-aminolevulinic acid and porphobilinogen nor the relative distribution of the urinary porphyrins. The role of the above substances in the induction of porphyria by hexachlorobenzene is discussed.

INTRODUCTION

The unexpectedly high rate of the biotransformation of hexachlorobenzene (HCB) has led us to find out whether its metabolites contribute to the effects exhibited by the parent compound. We had already observed that continuous administration of HCB leads to an accumulation of the major metabolites in the tissues of the treated animals[1]. It was of interest, therefore, to examine the effects of these substances. We tested pentachlorophenol, pentachlorothiophenol, and pentachlorothioanisol which was found to be formed from pentachlorothiophenol in the rat (Koss, unpublished results). Since Portig and coworkers observed that pentachlorothioanisol was converted to the 1-methyl(2,3,4,5,6-pentachlorophenyl)sulfoxide and to the 1-methyl(2,3,4,5,6-pentachlorophenyl)sulfone by rat liver microsomes (unpublished results) we also tested these substances.

MATERIAL AND METHODS

Animals and treatment

Groups of 4 to 5 female Wistar rats (Zentral-Institute für Versuchstierzucht, Hannover, Germany), weighing initially 150 g, received orally, by means of a stomach tube, pentachlorophenol (PCP), pentachlorothiophenol (PCThP), pentachlorothioanisol (PCTA), 1-methyl(2,3,4,5,6-pentachlorophenyl)sulfoxide (PCTA-0) and 1-methyl(2,3,4,5,6-pentachlorophenyl)sulfone (PCTA-0_2), respectively, dissolved in olive oil DAB 7 (5 %). This was administered every other day for several weeks. For the doses and further details consult table 1. Controls received oil under otherwise identical conditions. During the 24 hrs after the last dosing the animals were housed individually in metabolism cages for the separate collection of the excreta. The animals had access to Altromin food and tap water.

TABLE 1

TISSUE CONTENT OF CHLORINATED AROMATICS IN THE RAT AFTER LONG-TERM TREATMENT

Substance	Dose administered (µmoles/kg)	Dosing period (weeks)	Content of the compound in liver (nmoles/g)	adipose tissue (nmoles/g)
PCP	57[1]	4	75[2]	15
PCThP	400	7	2	70
PCTA	400	7	1	460
PCTA-0	150	7	n.d.[3]	n.d.
PCTA-0_2	150	7	n.d.	n.d.
HCB[4]	178	9	1000	36000

[1]PCP was administered daily (the other substances were administered every other day)

[2]Each value represents the mean of at least 3 animals.

[3]Not detected (the lower detection limit was determined to be 0.03 nmoles/g wet tissue).

[4]See Koss et al.[1]

SUBSTANCES

PCP (Fluka AG/Switzerland), m.w. = 264, was dissolved in 0.1 N NaOH. In order to eliminate contaminants (dioxines, dibenzofuranes) the solution was extracted with cyclohexane/diethyl ether 1:1 v/v. The aqueous phase was acidified with concentrated hydrochloric acid and extracted with diethyl ether. The ether was evaporated and the remaining PCP was recrystallised from chloroform/n-pentane 1:1 v/v; its melting point was measured to be 187 - 189.5 °C (2 : 190-191 °C). PCThP (Aldrich Chem. Co/USA), m.w. = 280, was recrystallised from n-hexane; its melting point was determined to be 231 °C (3 : 232-234 °C). PCTA was obtained after methylation of PCThP by reaction with diazomethane in etheral solution. Following recrystallization from n-hexane its melting point was 95 °C (4 : 95 - 96 °C). The sulfoxide (m.w. = 310) and sulfone (m.w. = 326) of PCTA were synthesized following the method of Beyer[5]. Their melting points were measured to be 162 - 163 °C and 193 °C, respectively. Tests by means of gas chromatography (GC) showed the substances to be of more than 99.5 % chemical purity.

ANALYTICAL PROCEDURES

For GC analysis the tissues and excreta were processed as described earlier[6]. Recovery of the substances was determined to be almost complete. For the determination of δ-aminolevulinic acid (ALA), porphobilinogen (PBG) and the porphyrins, from 0.1 to 0.5 ml of urine and 50 mg of liver tissue were sampled. The quantitative analysis of ALA and PBG was performed according to the method of Doss et al.[7], that of the urinary porphyrins according to Doss[8], and Doss et al.[9], and that of the liver porphyrins according to Seubert et al.[10].

RESULTS

The rats exposed to PCP, PCThP, PCTA, PCTA-O and PCTA-O_2 for several weeks (see details in table 1) did not differ from control animals in either appearance or behaviour. Their body weight gain and the weight of their liver, spleen and kidneys were also found not to be influenced by these substances (for the organ weights see table 2).

Analysis by means of GC showed PCP, PCThP and PCTA to be present in the liver and adipose tissue, while no measurable amounts of PCTA-O_2 were detected in these tissues (table 1).

Determination of the total liver porphyrins, and of ALA, PBG and total porphyrins in the urine revealed that with the exception of HCB none of the other substances led to measurable disturbances of the porphyrin pathway (table 3). Tests by means of thin layer chromatography showed the relative content of uro- and heptacarboxylic porphyrin in the liver and urine to be elevated and of the porphyrins with 6 to 2 carboxylic groups to be decreased only in the HCB-treated animals. However, the relative distribution of the liver and urinary porphyrins in the animals treated with the other substances did not differ from those in the control animals.

TABLE 2

RELATIVE ORGAN WEIGHT IN RATS AFTER LONG-TERM TREATMENT WITH CHLORINATED AROMATICS

Substance	Relative weight[1] of		
	Liver	Kidney	Spleen
Control	3.56 ± 0.28	0.66 ± 0.10	0.22 ± 0.01
PCP	- [2]	-	-
PCThP	3.65 ± 0.11	0.67 ± 0.04	0.23 ± 0.02
PCTA	3.62 ± 0.07	0.62 ± 0.10	0.21 ± 0.02
PCTA-O	4.08 ± 0.34	0.68 ± 0.02	0.34 ± 0.15
$PCTA\text{-}O_2$	3.86 ± 0.20	0.66 ± 0.05	0.27 ± 0.04
HCB[3]	5.80 ± 1.11	0.78 ± 0.04	-

[1]In per cent of the body weight

[2]Not determined

[3]See Koss et al.[1]
(Consult text and table 1 for more details).

TABLE 3

TOTAL LIVER PORPHYRINS AND THE CONTENT OF ALA, PBG AND TOTAL PORPHYRINS IN THE URINE OF RATS AFTER LONG-TERM TREATMENT WITH CHLORINATED AROMATICS

Substance	Total liver porphyrins (μg/g tissue)	Urinary ALA (μg/24 hrs)	Urinary PBG (μg/24 hrs)	Total urinary porphyrins (μg/24 hrs)
Control	0.4 ± 0.1[1]	16.9 ± 5.9	4.6 ± 1.9	6.9 ± 4.7
PCP	0.9 ± 0.6	12.0 ± 1.4	5.6 ± 1.5	1.8 ± 0.3
PCThP	0.4 ± 0.1	14.0 ± 9.1	9.6 ± 7.4	11.2 ± 9.3
PCTA	0.3 ± 0.1	12.5 ± 5.2	3.1 ± 1.0	4.8 ± 1.9
PCTA-O	0.4 ± 0.1	5.1 ± 1.2	4.6 ± 0.9	3.7 ± 1.5
$PCTA\text{-}O_2$	0.3 ± 0.1	10.7 ± 3.8	7.6 ± 2.7	4.7 ± 1.8
HCB[2]	10.3 ± 5.4	38.5 ± 11.4	36.2 ± 23.9	85.0 ± 4.5

[1]Figures are the mean ± S.D. of at least 3 animals.

[2]See Koss et al.[1]

(Consult text and table 1 for more details).

DISCUSSION

Since we know that HCB is metabolized to a relatively high extent, it is considered to be of importance for the elucidation of its toxic effects to study the fate of the metabolites identified so far. In the present experiment we observed that, in spite of the administration of relatively high doses of the substances tested, their concentration in the liver and adipose tissue was low or even not detectable. It is very probable that these metabolites of HCB are rapidly transformed to compounds with another chemical structure and excreted. The extent of their biotransformation was found to increase in the order PCP ⟶ PCTA ⟶ PCThP ⟶ PCTA-O = PCTA-O_2. In the case of HCB the rate of biotransformation amounted to about 50 %, while it was almost 100 % in the case of the sulfoxide and sulfone of PCTA (Koss et al., manuscript in preparation). It appears that the metabolites of HCB have a much lower persistance than the parent compound itself.

Contrary to our expectation the metabolites tested had no influence on body weight gain, on the weight gain of the liver, kidney and spleen and on the porphyrin pathway. With respect to the induction of porphyria it is of interest to note that Sinclair et al.[11] observed an accumulation of porphyrins in liver homogenates after application of PCP. This disturbances of porphyrin synthesis, however, was brought about only when ALA was added to the homogenate. In analogy with their finding we have shown that PCThP also disturbs the porphyrin metabolism only in the presence of a relatively high amount of ALA[12]. From these experiments the conclusion may be drawn that only those substances which are capable of inducing the ALA-synthase may lead to disturbances in the porphyrin pathway. With the exception of HCB none of the other substances tested, however, was found to increase the activity of this enzyme as revealed by the urinary ALA-content. A direct influence of the tested metabolites of HCB upon the synthesis of porphyrins can be therefore be excluded.

Nevertheless, it may be possible that the metabolites contribute indirectly to the toxic effects brought about by HCB. In a collaborative study (see this volume) evidence was obtained that compounds interfering with the glutathione system[13] can enhance the porphyrinogenic action of HCB. Since it is known that a conversion of HCB to sulfur-containing metabolites takes place, an interference of HCB itself with the sulfur-regimen in the cell is very likely to occur. Whether this conception contributes to the clearing-up of the porphyrinogenic action of HCB remains to be studied in future experiments.

ACKNOWLEDGEMENTS

Mrs. J. Seidel's skilful assistance is gratefully acknowledged.

REFERENCES

1. Koss, G., Seubert, S., Seubert, A., Koransky, W. and Ippen, H. Accepted for publication in Arch. Toxicol.
2. Residue Review, (1965) 10, 110.
3. Betts, J.J., James, S.P. and Thorpe, W.V. (1961) Biochem. J., 61, 611.
4. Tadros, W. and Saad, E. (1954) J. Chem. Soc., Part I, 1155.
5. Beyer, H. (1968) Lehrbuch der organischen Chemie, S. Hirzel Verlag, Leipzig, 15th. - 16th. ed., p. 116.
6. Koss, G., Koransky, W. and Steinbach, K. (1976) Arch. Toxicol., 35, 107.
7. Doss, M. and Schmidt, A. (1971) Z. Klin. Chem. Clin. Biochem., 9, 99.
8. Doss, M. (1970) Z. Klin. Chem. Klin. Biochem., 8, 197.
9. Doss, M. and Schmidt, A. (1971) Z. Klin. Chem. Klin. Biochem., 9, 415.
10. Seubert, S. and Seubert, A. (1977) Dtsch. med. Wschr., 102, 1882.
11. Sinclair, R. and Granick, S. (1974) Biochem. Biophys. Res. Communic. 61, 124.
12. Koss, G., Seubert, S., Seubert, A., Koransky, W. and Ippen, H. (1977) Abstract, presented at the Joint Meeting of Germany and Italian Pharmacologists, Venice, Oct. 4 - 6.
13. Kosower, E.M. (1976) in: Glutathione - metabolism and function, Raven Press, New York, 1 pp.

Chemical Porphyria in Man, J.J.T.W.A. Strik and J.H. Koeman eds.

AN APPROACH TO ELUCIDATE THE MECHANISM OF HEXACHLOROBENZENE-INDUCED HEPATIC PORPHYRIA, AS A MODEL FOR THE HEPATOTOXICITY OF POLYHALOGENATED AROMATIC COMPOUNDS (PHA's)

F.M.H. DEBETS AND J.J.T.W.A. STRIK
Department of Toxicology, Agricultural University, Wageningen, The Netherlands

Fig. 1. MECHANISM OF ACTION OF HEPATIC PORPHYRIA CAUSED BY POLYHALOGENATED AROMATIC COMPOUNDS. (PHA)

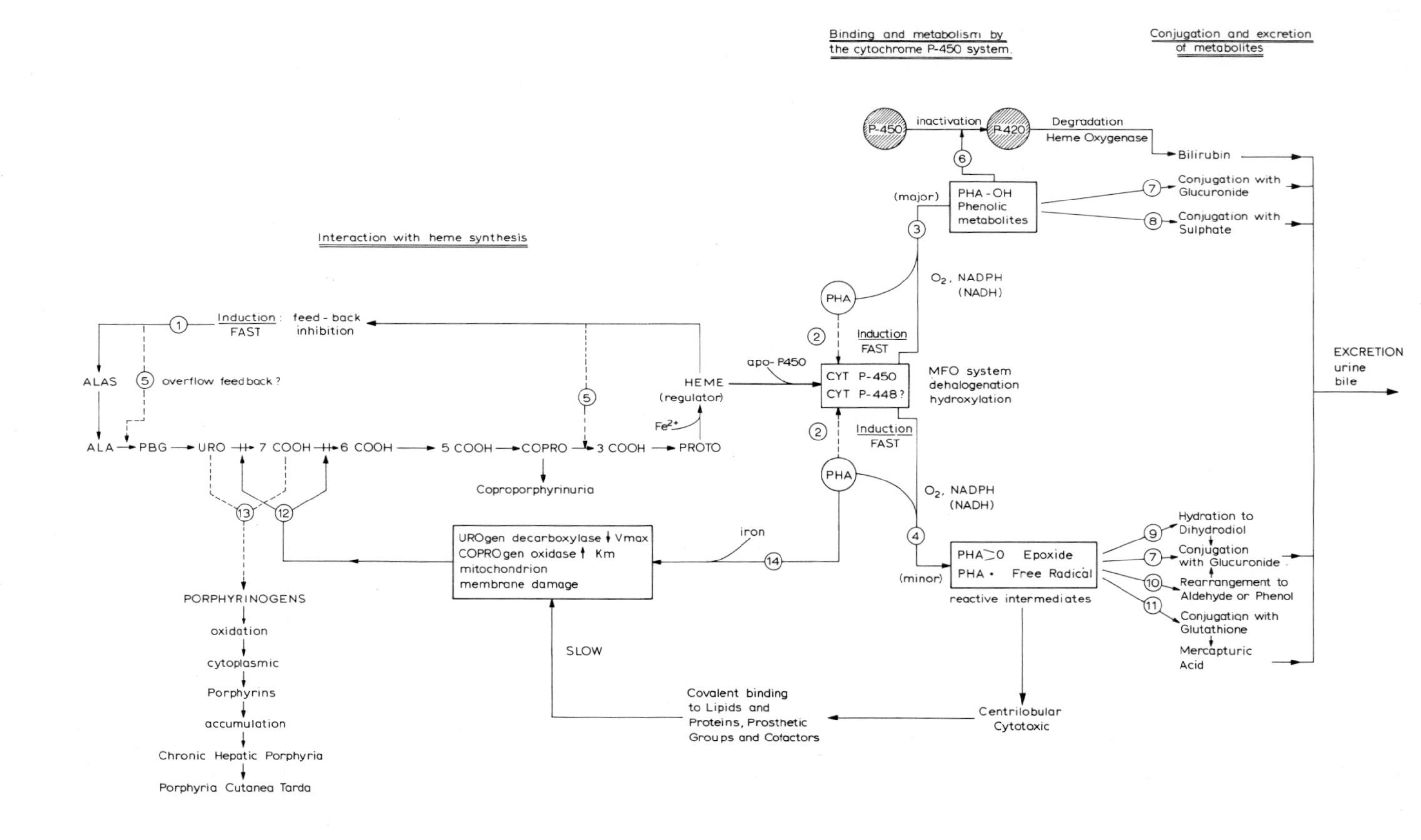

I. INTRODUCTION

The mechanism of the porphyrinogenic action of hexachlorobenzene (HCB) and other polyhalogenated aromatic hydrocarbons is still obscure. Elucidating the mechanism of hexachlorobenzene - induced porphyria may serve as a model for the hepatotoxic action of other polyhalogenated aromatic compounds (PHA's). Water-insolubility, lipophilic nature, planar or near-planar molecular structure, and different degree of halogenation are the common physicochemical properties of this group of compounds. In addition to their porphyrinogenic action[1-6], they induce microsomal drug-metabolizing enzymes[7-18] are nevertheless very slowly metabolized[19-26], and accumulate in animal[27,28] and human [29-32] adipose tissue. In the past much attention has been paid to the adaptive phase of microsomal drug-metabolizing enzyme induction[7-18] and the detection of morphological changes in the liver by the time porphyria develops[33-44]. Studies in recent years on the pharmacokinetics of PHA's[21,45-51] on the identification of metabolites[22,25,47,52-55] and their influence on porphyrin metabolism[56-59], and on the toxicity of the parent compounds[60], are promising. They provide further insight in the mechanism of action of PHA's. The theories that have been proposed to explain the porphyrinogenic action of polyhalogenated aromatic compounds will be reviewed and discussed with the help of the scheme presented in Figure 1. Special reference will be given to hexachlorobenzene, the member of this group of chemicals that has been most extensively studied as a porphyrinogenic agent and as a model for human symptomatic porphyria (also referred to as porphyria cutanea tarda, porphyria cutanea tarda symptomatica, chronic hepatic porphyria)[13,61].

II. INDUCTION OF δ-AMINOLEVULINIC ACID SYNTHASE (ALAS); INCREASE IN URINARY EXCRETION OF δ-AMINOLEVULINIC ACID (ALA) AND RELATIVE INCREASE IN URINARY COPROPORPHYRIN (see Fig. 1: no. 1, 5)

The induction of ALAS after chronic exposure to PHA's can be divided into two phases. The first one is the adaptive phase, characterized by an immediate slight (not greater than 2-3 fold) increase in hepatic ALAS activity after commencement of the treatment. This phase probably reflects a greater demand for heme due to the induction of microsomal hemoproteins (cytochrome P-450) by PHA's[3,6,8,11,13,62-67]. The second or pathological phase, marked by a 10-20 fold increase in ALAS activity, coincides with the onset of porphyrin accumulation in liver and increased urinary porphyrin excretion[68,69]. This second phase of ALAS induction is related to a decrease in the activity of uroporphyrinogen decarboxylase (UROG-D) and reflects a regulatory mechanism operating to keep the level of the free heme constant, in spite of inhibition of UROG-D (one of the enzymes in the pathway of heme synthesis, see VIII). Induction of ALAS and inhibition of UROG-D leads to an excessive urinary excretion of accumulating substrates of the defective enzyme, like amino-levulinic acid (ALA), porphobilinogen (PBG) and porphyrins with 7 and 8 carboxylic groups[46,69] (see Fig. 1). Induction of ALA-synthase or an elevated ALA-level seems to be a prerequisite for the onset of hepatic porphyria. Pentachlorophenol and pentachlorothiophenol, two metabolites of HCB, are not capable of inducing ALA-synthase[59] and they only disturb the heme synthesis in the presence of an elevated ALA concentration[59,70].

III. INDUCTION OF CYTOCHROME P-450 (P-448)-MEDIATED MIXED-FUNCTION OXIDASE ACTIVITY IN RELATION TO PORPHYRIA (see Fig. 1: no. 2)

Polyhalogenated aromatic compounds increase the concentration of cytochromes P-450 and b-5 and the activity of other components of the mixed-function oxidase system in the liver of mammals, birds and man[5,7-18,60,69,71-74]. In long-term experiments the increase in the activity of microsomal cytochrome P-450 and mixed-function oxidase (aminopyrine N-demethylation; ethoxyresorufin O-de-ethylation) reaches a maximum after 2 weeks exposure time[10,13,60]. The cytochrome P-450 concentration shows no drastic change when porphyria develops[13,60]. However, the activity of ethylmorphine N-demethylase, aminopyrine N-demethylase and ethoxyresorufin O-de-ethylase decreases upon chronic feeding of hexachlorobenzene, by the time porphyria develops[8,60]. Hexachlorobenzene [16,60] polychlorinated biphenyls[9,17,75] and polybrominated biphenyls[18,76,77] produce a mixed pattern of microsomal enzyme induction that shows characteristics of the 3-methylcholanthrene (P-448) and characteristics of the phenobarbital (P-450) classes of inducers. Stonard[17] suggests that there may exist a relationship between the mixed pattern of microsomal enzyme induction and the onset of hepatic porphyria. Metabolism of PHA's and accumulation of metabolites within the liver may be a prerequisite for the observed mixed pattern of enzyme induction and/or the lesion leads to the ultimate porphyric picture. These speculations contrast with the observation that TCDD - the most potent porphyrinogenic agent known - induces a pattern of microsomal enzymes identical to that caused by 3-methylcholanthrene[78-80]. Recently Jones and Sweeney[81] reported that mice genetically responsive to the induction of aryl hydrocarbon hydroxylase developed hepatic porphyria after treatment with 2,3,7,8-tetrachlorodibenzo-p-dioxin (TCDD). Nonresponsive mice, on the other hand, did not develop porphyria under the same conditions. Uroporphyrinogen decarboxylase activity was only depressed in the responsive mice. These observations suggest that induction of cytochrome P-448 and aryl hydrocarbon hydroxylase by PHA's are probably necessary for inhibition of uroporphyrinogen decarboxylase and hepatic porphyrin accumulation. Simultaneous administration of HCB and phenobarbital (an inducer of cytochrome P-450 and mixed-function oxygenases) promotes the metabolism of HCB and leads to an enhancement of the porphyrinogenic action of the former compound[82]. A metabolite or reactive intermediate of the parent compound is suspected to be responsible for the disturbance of hepatic porphyrin metabolism (see V and VIII).

IV. FORMATION OF PHENOLIC METABOLITES (see Fig. 1: no. 3 and Fig. 2)

Parke and Williams[19] were the first to study the fate of hexachlorobenzene in rabbits. They failed, however, to detect metabolites in the faeces and urine after a single oral dose of HCB suspended in water. A poor absorption of HCB from the gastro-intestinal tract was probably the cause of the fact that no metabolites were detected. Recent studies of the fate of HCB in rat[22,47,83-86], mouse, guinea pig, japanese quail, laying hen, rainbow trout[87], green sunfish (*Lepomis cyanellus*)[88], and monkey[89,90] have proved beyond any doubt that HCB in these species is metabolized into more polar compounds. In the rabbit, no phenolic metabolites of HCB have been found[91]. Iatropoulos *et al.*[92] showed that the major part of a single oral dose of HCB is slowly absorbed by the lymphatic system of the gastro-intestinal tract and deposited in adipose tissue, thus bypassing the portal venous transport to the liver. This may be an explanation for the slow metabolic conversion of HCB and the failure of several investigators to quantify appreciable amounts of metabolites in urine and faeces after a single dose of HCB. Pentachlorobenzene (PeCB), tetrachlorobenzene (TCB) and trichlorophenols (TCP) were only found after short-term administration of HCB to rats[47,83,86] and rhesus monkeys[90] (Fig. 2). The lower chlorinated benzenes may be derived partly from HCB as contaminants[47,93,94]. Pentachlorobenzene could not be detected in a long-term feeding experiment[46], which is not surprising because it has been demonstrated that pentachlorobenzene is almost completely biodegraded in the rat[95]. Pharmacokinetical studies of Koss *et al.*[46] showed that during long-term administration of HCB to rats about 60% of the xenobiotic was excreted unchanged and about 40% in the form of metabolites, once an equilibrium between intake and excretion was established. The ratio of the excreted phenolic metabolites (PCP and TCH; see Fig. 2) to the sulfur-containing metabolites (PCThP and TCThP; see Fig. 2) was 2.8 after long-term administration versus 1.3 after

Fig. 2. The metabolic fate of hexachlorobenzene HCB in the rat.

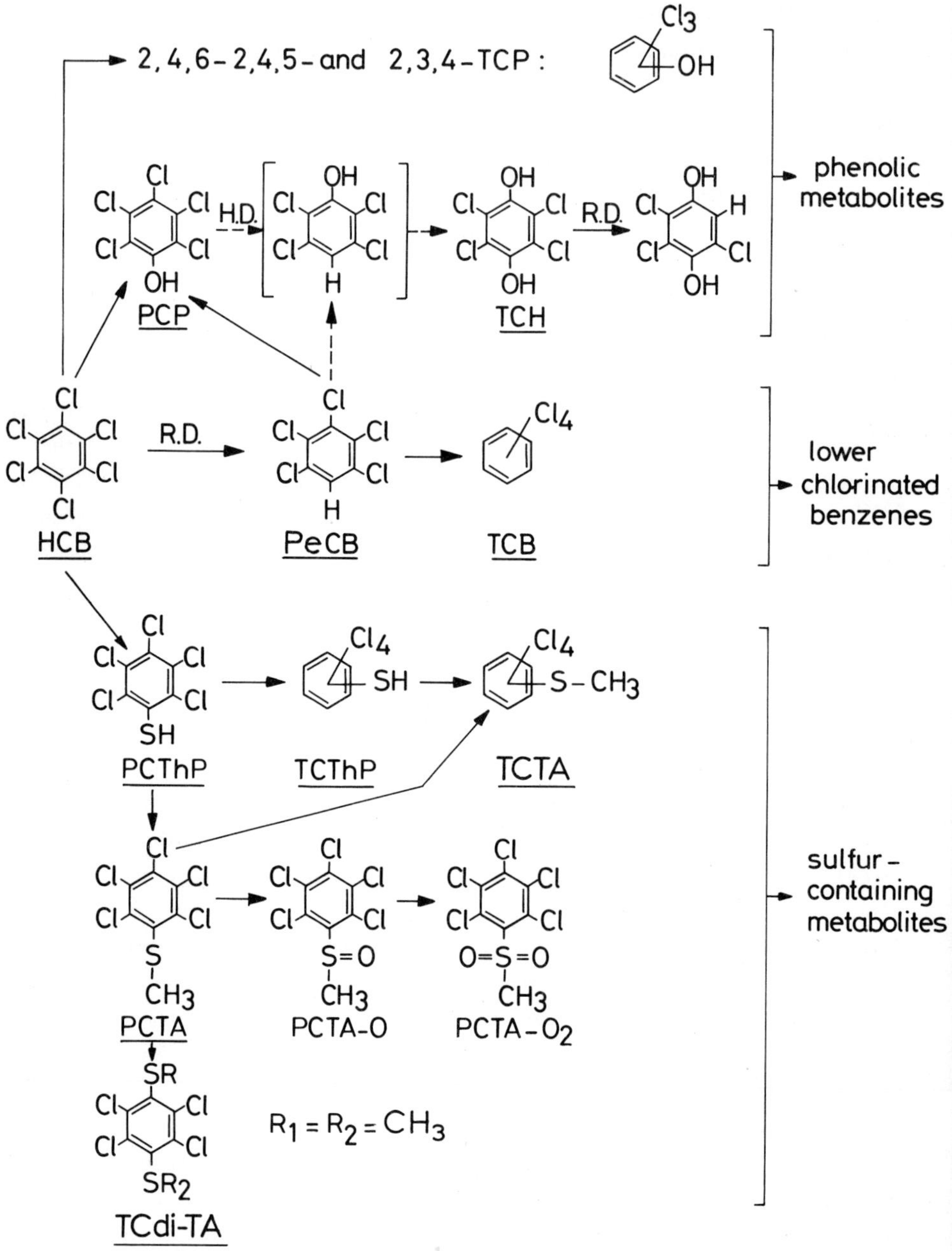

2,4,6- 2,4,5- and 2,3,4-TCP: Cl_3 –OH

H.D.

R.D.

PCP

TCH

phenolic metabolites

HCB

R.D.

PeCB

TCB

Cl_4

lower chlorinated benzenes

PCThP

TCThP

TCTA

Cl_4 SH

Cl_4 –S–CH_3

PCTA

PCTA-O

PCTA-O_2

sulfur-containing metabolites

TCdi-TA

$R_1 = R_2 = CH_3$

short-term administration of HCB[22,46]. In all animal species examined so far, with the exception of the japanese quail[87], pentachlorophenol was identified as the major metabolite of HCB[22,87,88]. Pentachlorophenol accounted for 45% of the total amount of metabolites recovered in the body, urine and faeces of rats treated with a single dose of HCB. Metabolism of hexachlorobenzene to pentachlorophenol is likely to proceed via hydrolytic dechlorination[96], as has been demonstrated in the metabolic conversion of 2,4,6-trichloroaniline to 4-amino-3,5-dichlorophenol.

Based on the observations that prolonged administration of pentachlorophenol to female rats induces no porphyria, several authors[58,75,94,97] conclude that pentachlorophenol as the major metabolite of HCB plays no prominent part in the disturbance of hepatic porphyrin metabolism. This argument is no longer valid in view of the recent findings that pentachlorophenol disturbs hepatic detoxification mechanisms[98] (see V) and accelerates the onset of HCB-induced hepatic porphyria[60]. In rats treated with a combination of HCB and PCP the onset of hepatic porphyria started about 3-4 weeks earlier than in rats treated with HCB alone (Debets and Strik, manuscript in preparation).

The metabolism of polychlorinated biphenyls (PCB's) has been thoroughly investigated by Hutzinger, Safe and co-workers and this subject has been reviewed recently[25]. As with hexachlorobenzene the major metabolites were phenolic compounds (see Ref. 25 and papers cited therein). For example studies on the metabolism of 4,4'-di-chlorobiphenyl in the rat demonstrated the formation of four monohydroxy-, four dihydroxy- and two trihydroxy metabolites[99]. Hydroxylated derivatives are also the main metabolic products of bromobiphenyls in the rabbit[100]. To what extent the formation of phenolic metabolites contribute to the porphyrinogenic effects and hepatotoxicity of polyhalogenated aromatic compounds remains to be investigated further.

V. DISTURBANCE OF MIXED-FUNCTION OXIDASE ACTIVITY AS A RESULT OF INACTIVATION OF MICROSOMAL CYTOCHROME P-450 BY PHENOLIC METABOLITES (see Fig. 1: no. 6)

Chlorophenols have been known for a long time as uncouplers of mitochondrial oxidative phosphorylation both *in vitro*[101,102] and *in vivo*[103]. Recently Arrhenius *et al.*[98] and Carlson[104] reported that phenolic compounds e.g. pentachlorophenol, di, tri- and tetrachlorophenols exert an inhibitory effect on another electron transport chain in the smooth endoplasmic reticulum of the cell, i.e. the microsomal detoxification enzyme chain. Pentachlorophenol strongly inhibited the electron transport between a flavin and cytochrome P-450 in the microsomal mixed-function oxygenase system *in vitro*. This resulted in a selective blocking of the cytochrome P-450 dependent C-oxygenation of aromatic amines, favouring their flavin mediated N-oxygenation[98]. This type of change in the detoxification pattern appeared to be associated with the formation of reactive electrophilic intermediates e.g. epoxides[105,106], which are supposed to be responsible for disturbing effect upon several cell functions, including mutagenic and carcinogenic effects by covalent binding to cellular macromolecules. Similar effects were obtained with various phenolic metabolites of polychlorinated biphenyls[98]. These results are consistent with our findings that the inhibition of the P-450 dependent O-demethylation of p-nitroanisole by pentachlorophenol in vitro is due to a selective inactivation of cytochrome P-450, which is converted to its metabolically inactive P-420 form[155] (Debets and Strik, manuscript in preparation). After the conversion of cytochrome P-450 to P-420, the heme prosthetic group of the latter can be oxidatively degraded by the microsomal heme oxygenase to form biliverdin[107] (Fig. 1). Biliverdin becomes subsequently reduced by biliverdin reductase to bilirubin, that can be excreted in bile and urine. Similar results were also obtained by Carlson, who found that 2,4,5-trichlorophenol at a dietary concentration of 400 mg/kg/day decreased hepatic microsomal cytochrome P-450 content. 2,3,5-, 2,3,6- and 2,4,6-trichlorophenol inhibited the demethylation of p-nitroanisole *in vitro*, when added to a final concentration of 0.25 mM[104]. The uncoupling effect of chlorophenols on the mitochondrial respiratory chain has been interpreted as a change in the properties of the lipid membrane, which carries the enzymes of the mitochondrial electron transport chain[108].

The same might be the case with the smooth endoplasmic reticulum membrane, which carries the enzymes of the mixed-function oxidase system. Indeed, it has been reported that compounds with a amphophilic character - i.e. having a hydrophilic and hydropholic (aromatic) part, like pentachlorophenol - readily form a complex with amphophilic phospholipids in cellular membranes[109,110], leading to an alteration in physicochemical properties of the lipid[111]. The observed inactivation of cytochrome P-450 after incubation of microsomes with PCP could therefore reflect a change in physicochemical properties of the lipid in which cytochrome P-450 is embedded. As a results of this, cytochrome P-450 might become detached from the microsomal membrane or its catalytic moiety might undergo a conformational change, losing thereby its metabolic functions.

Pentachlorophenol formed endogenously by metabolism of hexachlorobenzene, was demonstrated not only in the excreta, but in the liver as well[46]. Arrhenius *et al.*[112] showed that PCP administered *in vivo* accumulates markedly in the microsomal fraction of the liver, which encloses the drug-metabolizing system.

Pentachlorophenol generated in the smooth endoplasmic reticulum might alter the pattern of metabolism of the parent compound (HCB) or shorten the turnover rate of cytochrome P-450, as it accumulates *in situ*.

The results of several investigators discussed above, demonstrate clearly that phenolic metabolites of polyhalogenated aromatic compounds are able to disturb hepatic drug metabolism. The possibility that these phenolic metabolites play an essential part in the development of porphyria should therefore be considered seriously.

VI. FORMATION OF REACTIVE INTERMEDIATES AND SULFUR-CONTAINING METABOLITES (see Fig. 1: no. 4 and Fig. 2)

Another pathway of metabolism of polyhalogenated aromatics probably involving glutathione, leads to formation of sulfur-containing metabolites. Pentachlorothiophenol (PCThP), pentachlorothioanisol (PCTA), tetrachlorothiophenol (TCThP), tetrachlorothioanisol (TCTA) and tetrachlorodithioanisol could be demonstrated in the excreta of hexachlorobenzene-treated rats[22,47,55] (see Fig. 2). Pentachlorothiophenol and pentachlorothioanisol were also present in the livers of rats treated with hexachlorobenzene[47,55]. After short-term administration of HCB pentachlorothiophenol accounts for about 1/3 of the total excreted amount of metabolites[22]. The major part of the thiophenols could be isolated only after alkaline hydrolysis of the excreta. Free pentachlorothiophenol was, however, also extracted directly from the excreta[47,55]. A minor amount of this metabolite seems to be released from its conjugates with amino acids or proteins *in vivo*. Chasseaud[113] and Koss *et al.*[46] suggested that the conjugated compounds excreted in the bile could be hydrolyzed *in vivo* by the intestinal enzymes and/or microflora. The metabolites that are split off, may eventually be excreted in the faeces, or they may be reabsorbed and undergo enterohepatic circulation. Recently Larsen and Bakke[114] showed that an enzymic conversion of glutathione conjugates to methylthio derivatives takes place in the intestine, probably catalyzed by a C-S lyase of the intestinal flora or tissues.

The sulfur-containing metabolites can be partly derived from conjugation with glutathione. The recent detection and isolation of pentachlorophenylmercapturic acid in the urine of hexachlorobenzene-treated rats adds proof to this assumption[54]. Another considerable amount of sulfur-containing metabolites may stem from a reaction with methionine[52,55] or from covalent binding of reactive intermediates to functional sulfhydryl-groups of enzymes. Since HCB is metabolized slowly, escaping reactive intermediates may react readily with glutathione without lowering its concentration in the liver. This may explain why long-term administration of HCB to rats causes no appreciable decrease in hepatic glutathione levels[82,115,116]. Combined administration of HCB and diethylmaleate (a glutathion depleting agent), however, enhanced the porphyrinogenic action of HCB[82]. The elevated excretion of porphyrins and porphyrin precursors in rats chronically exposed to 1,2,4-trichlorobenzene could be reduced to almost normal levels by daily i.p. injections of glutathione, without stopping the treatment with 1,2,4-trichlorobenzene[115]. These results suggest that a toxic intermediate formed during HCB metabolism may be responsible for disturbing the heme synthesis. Metabolic hydrolytic dechlorination of HCB via highly reactive arene oxide intermediates does not seem to be the most plausible explanation, because of sterical hindrance of chlorine on the aromatic ring and the fact that no ortho-hydroxy derivatives of pentachlorophenol have been detected. Enzymatic epoxidation as an intermediate step in HCB metabolism must, however, not be ruled out completely, because investigations have shown that arene oxides are plausible intermediates in the metabolism of highly substituted chlorobenzenes like pentachlorobenzene, by the rabbit[91] and the rat[95] (see Fig. 3). Formation of free radicals during reductive dechlorination of HCB - as demonstrated by the metabolism of carbon tetrachloride to chloroform[117] - seems to be more probable, since the possibility of reductive dechlorination has been reported for hexachlorobenzene[83,118], chlorobiphenylols[119] and chlorophenols[120]. The chlorine atoms of HCB are labile against nucleophilic displacement, so that the possibility of a direct attack of glutathione on a carbon-chlorine bond of HCB-catalyzed by a glutathione chloro-transferase (see Fig. 3) - should be taken into consideration.

It has been well established that PCB's and PBB's are metabolized to hydroxylated derivatives via highly reactive arene oxide intermediates: compounds in which a formal aromatic double bound has undergone epoxidation via hepatic mono-oxygenase action[25,100,121] (Fig. 4). A low degree of chlorination of PCB's facilitates the formation of hydroxylated metabolites and, therefore, favours epoxidation. Sufur-containing metabolites of PCB's have been detected in the faeces and liver of mice injected once with 2,5,2',5'- or 2,4,2',4'-tetrachlorobiphenyl[52,53,122]. Two methylthio metabolites (3- and 4-methylthio-2,5,2',5'-TCB or 5- and 6-methylthio-2,4,2',4'-TCB) and two methylsulfone metabolites (at 3- and 4- or 5- and 6-position in the phenyl ring respectively) were excreted in the faeces (see Fig. 4). The total amount of the four sulfur-containing metabolites

Fig. 3. Possible metabolic pathways of pentachlorobenzene in the rat.

(G= glutathione)

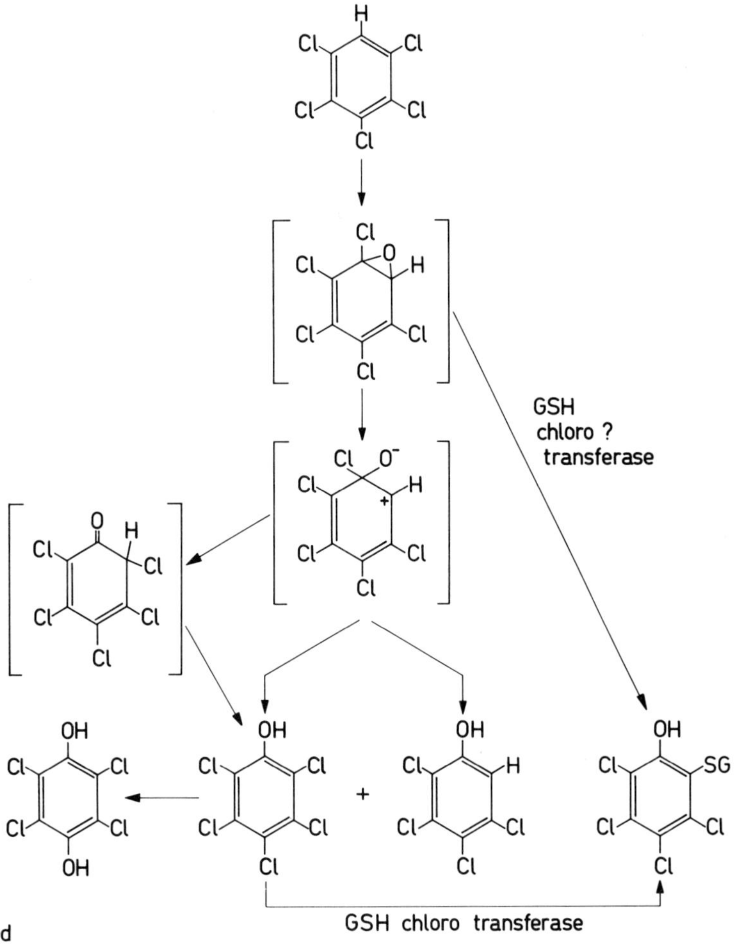

GSH chloro ? transferase

GSH chloro transferase

% excreted in					
feces		4.1	18	11.3	values from Koss [95]
urine	1.5	6.2	35	21.6	

Fig. 4. Metabolic pathways in the mammalian metabolism of halogenated aromatic hydrocarbons

(G= glutathione, M= $CH_2CH[(NHCOCH_3)(COOH)]$)

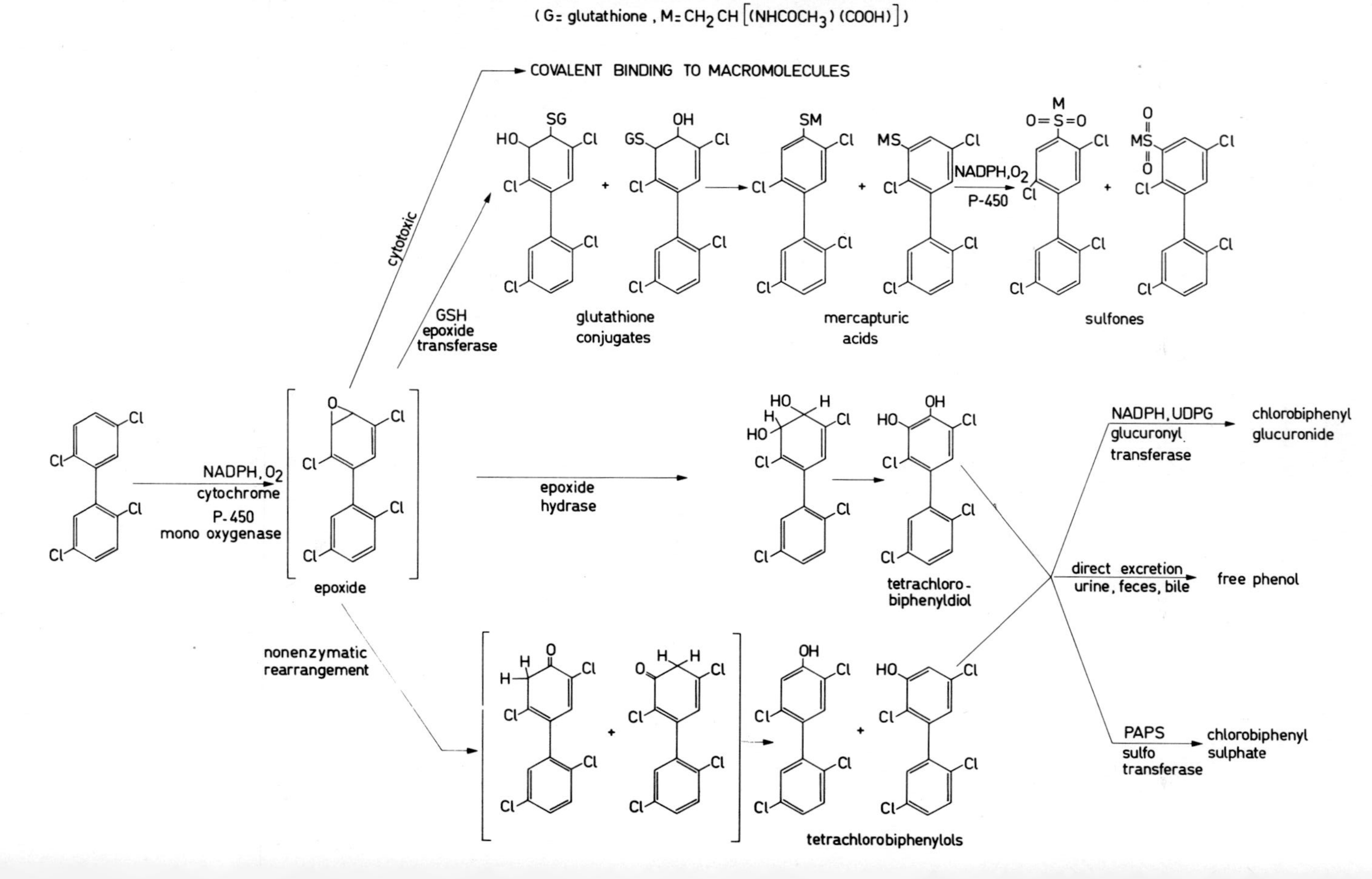

excreted during 6 days after administration of 2,4,2',4'-TCB accounted for only 0.12% of the dose[53]. The substitution by a methylthio or methylsulfone group at either of two adjacent positions in the phenyl ring points to an intermediate formation of arene oxides in the metabolism of these PCB isomers. In the liver only the methylsulfone metabolites could be detected and Mizutani[53] suggested that a rapid conversion of the methylthio metabolites to the corresponding methylsulfone metabolites by hepatic mono-oxygenase action could be responsible for this phenomenon. It should be noted that methylsulfone metabolites of PCB's were also found in seal blubber[123]. The biological and toxicological significance of sulfur-containing metabolites of halogenated aromatic hydrocarbons in relation to hepatic porphyria is still unknown. In only one instance the action of the HCB metabolite pentachlorothiophenol has been investigated in this respect. This sulfur-containing metabolite was found to be able to disturb hepatic porphyrin metabolism in the presence of an elevated ALA-level[124].

VII. INACTIVATION, CONJUGATION AND EXCRETION OF REACTIVE INTERMEDIATES AND METABOLITES (see Fig. 1: no. 7-11 and Fig. 4)

Both hexachlorobenzene[60,69] and polychlorinated biphenyls[44,75] increase the activity of hepatic glucuronyl transferase. This enzyme catalyses the conjugation of phenolic metabolites with glucuronic acid (see Fig. 4) to form glucuronides, that can be excreted easier in bile and urine. Induction of glucuronyl transferase suggests that phenolic metabolites of PHA's are excreted partly as glucuronides. Data, concerning the question to what extent phenolic metabolites are excreted free or as conjugates, are very limited. Exogenously administered PCP (the major metabolite of HCB) is excreted for the most part in the urine. About 60% of the amount of PCP excreted in the urine was found to be present as conjugated PCP in both mice[103] and rats[47,125]. Conjugation with glutathione and subsequent mercapturic acid formation, as a mechanism to inactivate the highly reactive epoxides or free radicals arising during the metabolism of PHA's (see VI), has become more plausible since the detection of sulfur-containing metabolites of PCB's (see Fig. 4) and the isolation of pentachlorophenylmercapturic acid in the urine of hexachlorobenzene-treated rats[54]. Polybrominated biphenyls[18] and probably also polychlorinated biphenyls induce hepatic epoxide hydratase, another enzyme involved in the inactivation of electrophilically reactive intermediates of PHA metabolism. Epoxide hydratase is probably localized in the endoplasmic reticulum, forming a complex with the mono-oxygenase system[126]. It catalyses the hydration of expoxides to the electrophilically unreactive dihydro-diols[127] (see Fig. 4). Rearrangement to a phenol is an alternative nonenzymatic pathway of epoxide inactivation. It is often accompanied by an NIH shift of a chlorine substituent and/or loss of a chlorine atom from an epoxide/keto-enol tautomer during the rearrangement to the phenol[121] (Fig. 3). A decrease in the activity of extrahepatic epoxide hydratase after treatment with polychlorinated biphenyls has also been reported: pregnant rats exposed to PBB's showed an increased mammary aryl hydrocarbon hydroxylase (AHH) activity, whereas epoxide hydratase activity was depressed to half the control activity[128,129]. In a long-term experiment with male rats, renal AHH activity was elevated to ten times the control level, whereas epoxide hydratase activity represented only about ten percent of its control activity[130].

The examples mentioned above show clearly that further investigations into the influence of halogenated aromatic hydrocarbons on the pathways involved in the inactivation, conjugation and excretion of their metabolites, are required for a complete evaluation of the toxicological properties and environmental hazards of this group of compounds.

VIII. INHIBITION OR INACTIVATION OF UROPORPHYRINOGEN DECARBOXYLASE BY POLYHALOGENATED AROMATICS OR THEIR METABOLITES (see Fig. 1: no. 12, 13, 14)

Prolonged exposure to polyhalogenated aromatic compounds leads in several animal species to the development of porphyria, which is marked by a progressive accumulation of porphyrins with 7 and 8 carboxylic groups and, to a lesser extent, porphyrins with 6, 5 and 4 carboxylic groups in the liver and urine[61,131-133]. Ockner[35] and De Matteis *et al.*[134] were the first to note this massive excretion of higher carboxylated porphyrins in HCB-porphyria and the latter suggested that this might be due to a defect in the uroporphyrinogen decarboxylating mechanism. Similar observations have been made with PCB's[3,44,70] and PBB's[67]. Recently it has been shown conclusively that the activity of hepatic uroporphyrinogen decarboxylase (UROG-D) - involved in the stepwise decarboxylation of uroporphyrinogen to coproporphyrinogen - is markedly decreased in HCB porphyria[62,135-139]. The same seems to be the case with other porphyrinogenic polyhalogenated aromatic compounds, because they induce an identical pattern of porphyrin accumulation[11]. The decrease in the activity of uroporphyrinogen decarboxylase in the course of chronic HCB administration to rats, coincided with a progressive accumulation and excretion of porphyrins with 8-5 carboxylic groups as porphyria developed[132,137]. These results suggest that chronic hepatic porphyria induced by PHA's can be attributed to inhibition or inactivation of hepatic uroporphyrinogen decarboxylase. Accumulation of uroporphyrin was prevented by treatment with SKF 525-A and piperonyl butoxide, two inhibitors of hepatic drug metabolism[70]. The porphyrinogenic action of hexachlorobenzene in female rats could be enhanced by promoting its metabolism with phenobarbital[82]. Co-administration of HCB with diethylmaleate, a glutathione depleting agent, also led to an enhancement of the porphyrinogenic action[82]. This and the isolation of sulfur-containing metabolites from the urine of rats treated with HCB and PCB's (see VI), strongly suggests that an electrophilic metabolic product of PHA's reacts with sulfhydryl-groups. The catalytic part of hepatic UROG-D, which most probably contains also sulfhydryl groups[140], could be inactivated through a reaction with electrophilic metabolites of PHA's.

These experiments provide indirect evidence supporting the hypothesis that a direct attack of a reactive intermediate or metabolite of the parent compound on uroporphyrinogen decarboxylase is responsible for disturbing the heme synthesis. Earlier speculations of Simon *et al.*[141] that the porphyrinogenic action might be attributable to HCB itself, perhaps by altering the membrane permeability after its incorporation into the membrane phospholipid layer, seem to be no longer valid in the light of these new facts.

IX. MORPHOLOGICAL CHANGES AND HEPATIC PORPHYRIA INDUCED BY POLYHALOGENATED AROMATIC COMPOUNDS

The morphological alterations in the liver of experimental animals fed with subchronic porphyrinogenic doses of HCB, PCB's and PBB's are similar[33-44,145]. A marked enlargement of hepatocytes, especially in the centrilobular part of the liver acini, can be already observed several weeks before porphyria develops. Electron microscopic examination shows that this liver cell hypertrophy is due to a considerable proliferation of the smooth endoplasmic reticulum (SER), as a result of drug metabolizing enzyme induction. The mitochondria are often dislocated into scattered clumps, in consequence of SER proliferation. Most mitochondria appear normal, but sometimes swollen mitochondria can be observed. Centrilobular hepatocytes frequently contain lipid vacuoles. Local small spots of centrilobular necrosis[8,66,134,143] or an increase in the plasma concentration of liver enzymes, indicating liver cell damage, are observed in several studies by the time porphyria develops[65,142].

The occurrence of eosinophilic concentric membrane arrays (synonyms in current use: whorls; myelin figures; inclusion bodies; hyalin bodies) in the cytoplasm of the enlarged hepatocytes is one of the most striking features of

HCB[5,38,41-43] and PCB[44,144] treatment. Whorls consist of concentric layers of tightly packed membranes, probably originating from SER of which the intracisternal cavity is collapsed. They frequently enclose a central core of lipid droplets[146] (Fig. 5). Long-term feeding of HCB to male rats results in the formation of large whorls, varying in size from 2 to 20 μm in diameter[38] and visible in the light microscope (Fig. 5). Females are more susceptible to the porphyrinogenic action of HCB[5,12,13,43] and PCB's[37] but develop only small whorls ranging from about 0.2 to 3.0 μm in diameter[60] (Fig. 6). The capacity to induce the formation of whorls is not restricted to polyhalogenated aromatic compounds. Numerous drugs and other lipophilic xenobiotics, like carbon tetrachloride[147], phenobarbital[148], DDT[149], dieldrin[150], acetaminoflu-rene[151], tetrasul[152], are known to evoke whorls in experimental animals.

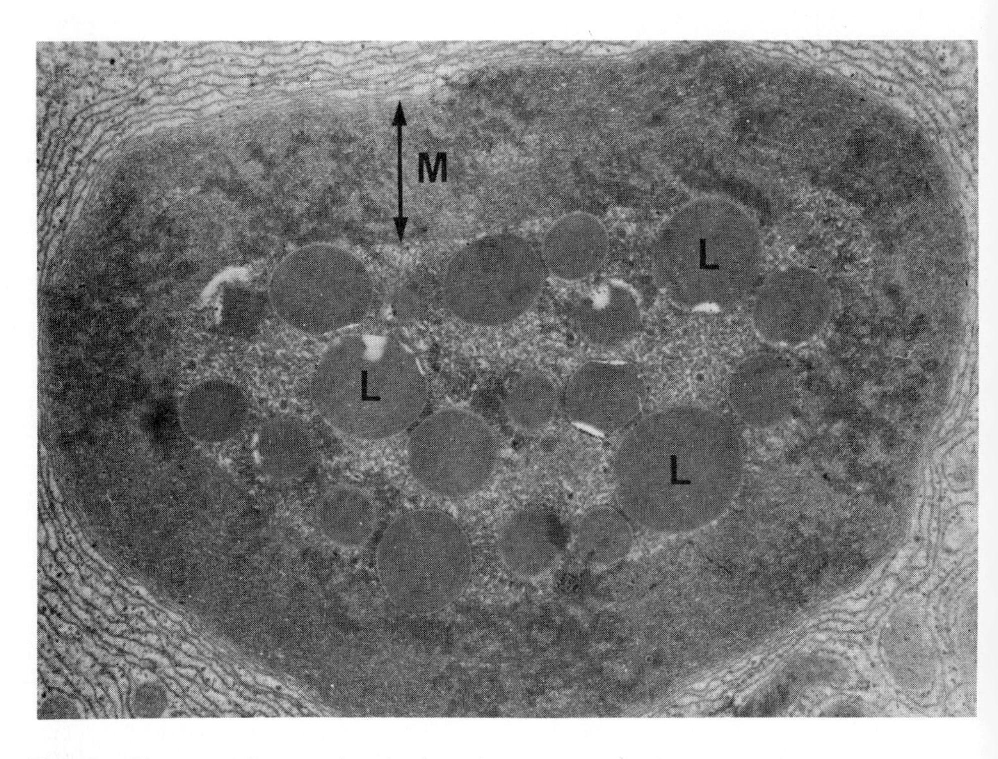

Fig. 5. Electron micrograph, showing the appearance of a large whorl in the cytoplasm of a hepatocyte from a male Wistar rat after subchronic feeding of 300 ppm HCB for 12 weeks. The whorl consists of a central core of lipid droplets (L), surrounded by a thick layer of concentric arrays of tightly packed smooth membranes (M). x 10,000.

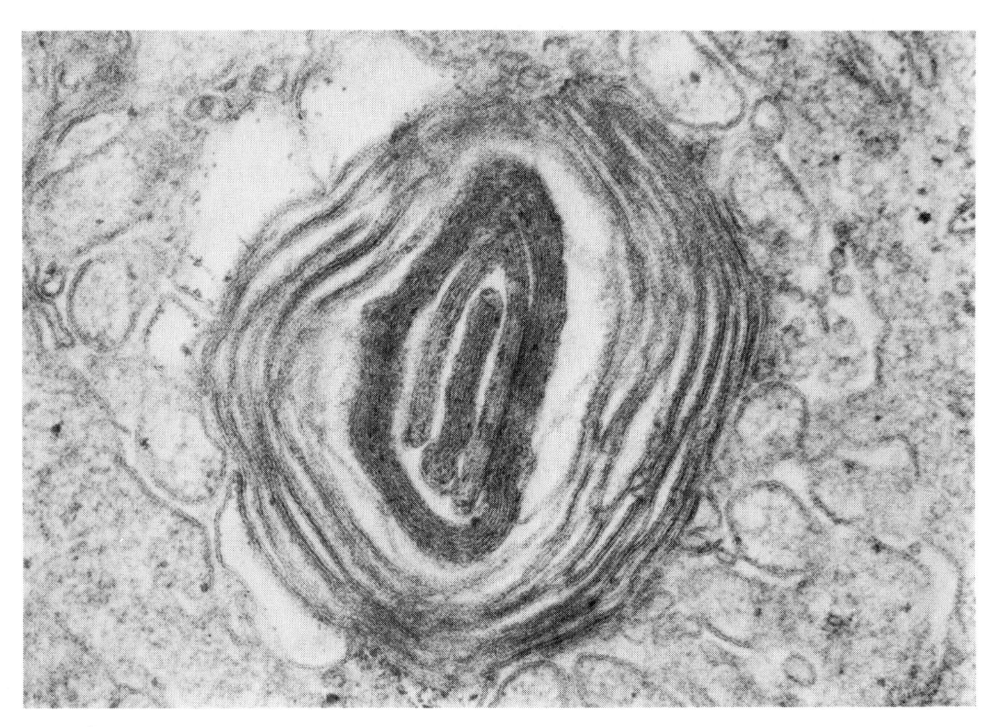

Fig. 6. Electron micrograph, showing the appearance of a small whorl in the cytoplasm of a hepatocyte from a female Wistar rat after subchronic feeding of 1000 ppm HCB for 8 weeks. x 100,000.

The function of whorl formation and its significance in relation to the development of porphyria is still obscure. However, since many whorls appear by the time porphyria develops[38,60], an interrelationship seems likely. Whorl formation may be involved in the elimination of fat-stored HCB or other lipophilic xenobiotics[41]. It seems reasonable to suggest that the SER starts to proliferate around lipid droplets, probably containing the lipophilic polyhalogenated aromatic compound. In an attempt to metabolize the lipophilic compound into more polar products, SER membranes can be damaged by formation of toxic metabolites and/or reactive intermediates (*e.g.* chlorophenols, arene oxides, free radicals; see IV, V, VI. In order to continue the metabolism, new SER membranes are concentrated around the damaged membranes enclosing the central lipid core. A whorl thus is conceived of as a cluster of inactive or hypoactive SER membranes enclosing lipid and at the periphery surrounded by a few intact concentric membrane arrays. The formation of whorls is, moreover, accompanied by a decrease in drug metabolizing enzyme activity in livers of HCB-exposed rats[42,60]. This observation suggests that the SER becomes hypoactive or even degenerative because toxic metabolites of HCB (*e.g.* pentachlorophenol) are continuously formed and subsequently accumulated up to a critical concentration *in situ* (see IV and V). The mechanism of whorl formation may be comparable with the formation of myeloid bodies, that can be observed in many cell types after administration of amphophilic drugs. Myeloid bodies are storage bodies containing membranes that cannot be further digested because of drug action. They should be distinguished from whorls, because they are surrounded by a lysosomal membrane and have no lipid core. This mechanism presupposes that amphophilic drugs readily form a complex with membrane phospholipids, leading to a change in the physico-chemical properties of these phospholipids. Normally, altered membranes are removed by autophagy. In contrast the drug-lipid complexes impair their own degradation in phagosomes and in lysosomes and are transformed into myeloid bodies (see ref. 111 and papers cited therein). Amphophilic metabolites of PHA's, like pentachlorophenol may react in a similar way with phospholipids of cellular membranes (see V). Amphophilic metabolites of PHA's generated in the SER by metabolism of the parent compound may directly form a complex with phospholipids of SER membranes. This not only causes inactivation of cytochrome P-450 (see V), and disturbance of liver mixed function oxygenase activity, but also induces whorl formation.

X. DELAYED ONSET OF HEPATIC PORPHYRIA: A PROPERTY INHERENT TO POLYHALOGENATED AROMATIC COMPOUNDS

Prolonged administration of PHA's to susceptible animal species leads to the development of hepatic porphyria. However, the time required for the appearance of symptoms may vary from a few weeks to one year, depending on the animal species and the administered dose. As yet no explanation for the delayed onset of this type of experimental porphyria has been found. Elder[153] suggests that the delay may be due, at least in part, to the time required to reach a critical porphyrinogenic concentration of the parent compound HCB in the liver. In this connection he refers to the work of Grant *et al.* and Vos *et al.* who showed that porphyria does not develop in female rats[12] and Japanese quail[154] until liver HCB concentrations have reached a certain value (30 µg/g liver for female rats). However, the rate of accumulation of HCB within the liver of female rats is the same as in males[12]. Nonetheless male animals are practically resistant to the porphyrinogenic action of HCB and administration of female sex hormones makes them more sensitive[164]. The response to HCB in females is decreased by ovariectomy and enhanced by oestrogens[165,166]. The role of gonadal steroid hormones in relation to the development of hepatic porphyria and the formation of large whorls in the cytoplasm of liver cells (chapter IX), is still obscure. The hypothesis of a critical porphyrinogenic concentration of HCB in the liver is unlikely for another reason: it has been shown that stimulation of HCB metabolism with phenobarbital accelerates the onset of hepatic porphyria[82]. These results are consistent with earlier suggestions of Stonard[17] that the slow onset of the mixed pattern of enzyme induction, after chronic administration of PHA's, may be related to the slow onset of hepatic porphyria (see III).

A more likely explanation for the delayed onset of hepatic porphyria is that a critical concentration of one of the metabolites is needed to cause damage to intracellular membranes and enzymes[60,155]. This critical level of a given metabolite might, moreover, alter the detoxification pattern of HCB in a direction that favours the formation of reactive electrophilic intermediates. The latter phenomenon has already been observed for the influence of chlorophenols on the metabolism of aromatic amines[98,105]. The reactive intermediate may be preferentially conjugated with glutathione, thereby depleting glutathione from the liver. Liver necrosis, alkylation of cellular macromolecules and porphyria occur when glutathione is no longer available. This may explain the concomitant delayed appearance of centrilobular liver necrosis and porphyrin fluorescence in HCB porphyria.

XI. SUMMARY AND CONCLUSIONS

Polyhalogenated aromatic compounds (PHA's) seem to have a common mechanism to affect the porphyrin metabolism. Increase in the urinary excretion of uroporphyrin and heptacarboxylic porphyrin in chronic hepatic porphyria induced by PHA's, is accompagnied by a decrease in the activity of hepatic uroporphyrinogen decarboxylase (UROG-D). These results lead to the conclusion that the decarboxylation of uroporphyrin is blocked by inhibition or inactivation of this decarboxylase. It has been shown conclusively that the porphyrinogenic effect of PHA's is not due to a direct interaction with the heme synthesis. A metabolite or reactive intermediate formed by metabolism of the PHA is suspected to be responsible for the disturbance of hepatic heme synthesis. PCB's, PBB's and possibly also HCB are metabolised via highly reactive arene oxide intermediates. The detection of sulfur-containing metabolites in the excreta of animals treated with PCB's or HCB suggests that these unstable intermediates may react with catalytic SH-groups of enzymes.

The main metabolites of PHA's, however, are the hydroxylated derivatives, which have been found to disturb hepatic drug metabolism *in vitro* by a selective inactivation of cytochrome P-450. Damaged SER membranes, carrying cytochrome P-450, seem to be removed by formation of whorls. It remains to be investigated what metabolites actually are involved in the onset of hepatic porphyria. The hypothesis that metabolism of PHA's is a prerequisite for the development of experimental porphyria does not agree with the observations that female rats are more susceptible to the porphyrinogenic effects of PHA's than males, though the latter show a more marked microsomal drug-metabolizing enzyme induction with these compounds. The delayed onset of hepatic porphyria might be explained in terms of time required to build up a critical porphyrinogenic concentration of a given metabolite in the liver or, as a result of this, by a change in the pattern of metabolism of the PHA. The influence of PHA's or their metabolites on the activity of enzymes involved in the inactivation or conjugation of electrophilically reactive intermediates, needs to be investigated further, because several cases in which a depressed activity was found have been reported. Additional studies on the lower chlorinated benzenes have to be carried out, because contradictory results have been published concerning their ability to induce hepatic porphyria[115,156]. Experiments with laboratory animals, carried out to gain more insight into the mechanism of chronic hepatic porphyria caused by PHA's, are time consuming due to the long period of administration required to evoke porphyria. The use of tissue culture of primary liver cells, acting as a fast reacting test system[64,70,157-163], may offer a more simple and successful alternative.

LIST OF ABBREVIATIONS

AHH, aryl hydrocarbon hydroxylase
ALA, δ-aminolevulinic acid
ALAS, δ-aminolevulinic acid synthase
COPRO, coproporphyrin
GSH, reduced glutathione
HCB, hexachlorobenzene
H.D., hydrolytic dechlorination
MFO, mixed function oxidase
PAPS, 3'-phosphoadenylsulphate
PBB's, polybrominated biphenyls
PBG, porphobilinogen
PCB's, polychlorinated biphenyls
PCP, pentachlorophenol
PCTA, pentachlorothioanisol
PCTA-O, sulfoxide of pentachlorothioanisol
$PCTA-O_2$, sulfone of pentachlorothioanisol
PCThP, pentachlorothiophenol
PeCB, pentachlorobenzene
PHA, polyhalogenated aromatic compound
PROTO, protoporphyrin
R.D., reductive dechlorination
SER, smooth endoplasmic reticulum
SKF 525A, β-diethylaminoethyldiphenylpropylacetate
TCB, tetrachlorobenzene
TCDD, 2,3,7,8-tetrachlorodibenzo-p-dioxin
TCH, tetrachlorohydroquinone
TCP, trichlorophenol
TCTA, tetrachlorothioanisol
TCdi-TA, tetrachlorodithioanisol.
TCThP, tetrachlorothiophenol
UDPG, uridine diphosphate-D-glucuronate
URO, uroporphyrin
UROG-D, uroporphyrinogen decarboxylase

REFERENCES

1. De Matteis, F. (1967) Pharmacol. Rev., 19, 523.
2. Vos, J.G. and Koeman, J.H. (1970) Toxicol. Appl. Pharmacol., 17, 656.
3. Goldstein, J.A., Hickman, P. and Jue, D.L. (1974) Toxicol. Appl. Pharmacol., 27, 437.
4. Strik, J.J.T.W.A. (1973) Meded. Rijksfac. Landbouwwetensch. Gent, 38, 709.
5. Strik, J.J.T.W.A. and Koeman, J.H. (1976) in Porphyrins in Human Diseases, Doss, M. (ed.), Karger, Basel, p. 418.
6. Goldstein, J.A., Hickman, P., Bergman, H. and Vos, J.G. (1973) Res. Commun. Chem. Path. Pharmacol., 6, 919.
7. Wada, O., Yano, Y., Urata, G. and Nakao, K. (1968) Biochem. Pharmacol., 17, 595.
8. Sweeney, G.D., Janigan, D., Mayman, D. and Lai, H. (1971) S. Afr. J. Lab. Clin. Med., 17, 68.
9. Alvares, A.P., Bickers, D.R. and Kappas, A. (1973) Proc. Nat. Acad. Sci. (Wash.), 70, 1321.
10. Rajamanickam, C., Amrutavalli, J., Rao, M.R.S. and Padmanaban, G. (1972) Biochem. J., 129, 381.
11. Strik, J.J.T.W.A. (1973) Enzyme, 16, 224.
12. Grant, D.L., Iverson, F., Hatina, G.V. and Villeneuve, D.C. (1974) Environ. Physiol. Biochem., 4, 159.
13. Stonard, M.D. (1974) Brit. J. Haemat., 27, 617.
14. Stonard, M.D. and Nenov, P.Z. (1974) Biochem. Pharmacol., 23, 2175.
15. Turner, J.C. and Green, R.S. (1974) Biochem. Pharmacol., 23, 2387.
16. Stonard, M.D. (1975) Biochem. Pharmacol., 24, 1959.
17. Stonard, M.D. and Greig, J.B. (1976) Chem.-Biol.-Interactions, 15, 365.
18. Dent, J.G., Netter, K.J. and Gibson, J.E. (1976) Toxicol. Appl. Pharmacol., 38, 237.
19. Parke, D.V. and Williams, R.T. (1960) Biochem. J., 74, 5.
20. Villeneuve, D.C. (1975) Toxicol. Appl. Pharmacol., 31, 313.
21. Koss, G. and Koransky, W. (1975) Arch. Toxicol., 34, 203.
22. Koss, G., Koransky, W. and Steinbach, K. (1976) Arch. Toxicol., 35, 107.
23. Kimbrough, R.D. (1974) CRC Crit. Rev. Toxicol., 2, 445.
24. Fishbein, L. (1974) Ann. Rev. Pharmacol., 14, 139.
25. Sundström, G., Hutzinger, O. and Safe, S. (1976) Chemosphere, 5, 267.
26. Vinopal, J.N. and Casida, J.E. (1973) Arch. Environ. Contam. Toxicol., 1, 122.
27. Koeman, J.H., Noever de Brauw, M.C. and Vos, R.H. de (1969) Nature (Lond.), 221, 1126.
28. Koss, G. and Manz, D. (1976) Bull. Environ. Contam. Toxicol., 15, 189.
29. Acker, L. and Schulte, E. (1970) Naturwiss., 57, 497.
30. Brady, M.N. and Siyali, D.S. (1972) Med. J. Aust., 1, 158.

31. Curley, A., Burse, V.W., Jennings, R.W., Villaneuva, E.C., Tomatis, L. and Akazaki, K. (1973) Nature, 242, 338.

32. Hammond, A.L. (1972) Science, 175, 155.

33. Bennett, G.A., Drinker, C.K. and Warren, M.F. (1938) J. Industr. Hyg., 20, 97.

34. Miller, J.W. (1944) Publ. Hlth. Rep. (Wash.), 59, 1085.

35. Ockner, R.K. and Schmid, R. (1961) Nature (Lond.), 189, 499.

36. Sweeney, G.D., Janigan, D., Mayman, D. and Lai, H. (1971) S. Afr. J. Lab. Clin. Med., 17, 68.

37. Kimbrough, R.D., Linder, R.E. and Gaines, T.B. (1972) Arch. Environm. Hlth., 25, 354.

38. Medline, A., Bain, E., Menon, A.I. and Haberman, H.F. (1973) Arch. Path. (Chic.), 96, 61.

39. Timme, A.H. Taljaard, J.J.F., Shanley, B.C. and Joubert, S.M. (1974) S. Afr. Med. J., 48, 1833.

40. Mollenhauer, H.H., Johnson, J.H., Younger, R.L. and Clark, D.E. (1975) Am. J. Vet. Res., 36, 1777.

41. Mollenhauer, H.H., Johnson, J.H., Younger, R.L. and Clark, D.E. (1976) Am. J. Vet. Res., 37, 847.

42. Kuiper-Goodman, T., Krewski, D., Combley, H., Doran, M. and Grant, D.L. (1976) in Proceedings of the Fourth International Congress of Stereology, National Bureau of Standards Special Publication 431, p. 351.

43. Kuiper-Goodman, T., Grant, D.L., Moodie, C.A., Korsrud, G.O. and Munro, I.C. (1977) Toxicol. Appl. Pharmacol., 40, 529.

44. Schmoldt, A., Altenähr, E., Grote, W., Dammann, H.G., Sidau, B. and Benthe, H.F. (1977) Arch. Toxicol., 37, 203.

45. Piper, W.N., Rose, J.Q. and Gehring, P.J. (1973) Environ. Hlth. Perspect., 5, 241.

46. Koss, G., Seubert, S., Seubert, A., Koransky, W. and Ippen, H. (1978) Arch. Toxicol., 40, 285.

47. Jansson, B. and Bergmann, A. (1978) Chemosphere, 3, 257.

48. Matthews, H.B. and Anderson, M.W. (1975) Drug. Metab. Dispos., 3, 211.

49. Matthews, H.B. and Anderson, M.W. (1975) Drug. Metab. Dispos., 3, 371.

50. Mehendale, H.M. (1976) Drug Metab. Dispos., 4, 124.

51. Peterson, R.E., Seymour, J.L. and Allen, J.R. (1976) Toxicol. Appl. Pharmacol., 38, 609.

52. Mio, T., Sumino, K. and Mizutani, T. (1976) Chem. Pharm. Bull. (Tokyo), 24, 1958.

53. Mizutani, T. (1978) Bull. Environm. Contam. Toxicol., 20, 219.

54. Renner, G., Richter, E. and Schuster, K.P. (1978) Chemosphere, 8, 663.

55. Koss, G., Koransky, W. and Steinbach, K. (1979) Arch. Toxicol., 42, 19.

56. Lui, H., Sampson, R. and Sweeney, G.D. (1976) in Porphyrins in Human Diseases, Doss, M. (ed.), Karger, Basel, p. 405.

57. Goldstein, J.A., Friesen, M., Linder, R.E., Hickman, P., Hass, J.R. and Bergman, H. (1977) Biochem. Pharmacol., 26, 1549.

58. Kimbrough, R.D. and Linder, E. (1978) Toxicol. Appl. Pharmacol., 46, 151.

59. Koss, G., Seubert, S., Seubert, A., Koransky, W., Ippen, H. and Straub, M., this volume.

60. Debets, F.M.H. and Strik, J.J.T.W.A., manuscript in preparation.

61. Doss, M., Schermuly, E. and Koss, G. (1976) Ann. Clin. Res., 8, Suppl. 17, 171.

62. Taljaard, J.J.F., Shanley, B.C. and Joubert, S.M. (1971) Life Sci., 10, 887.

63. Taljaard, J.J.F., Shanley, B.C., Deppe, W.M. and Joubert, S.M. (1972) Brit. J. Haemat., 23, 513.

64. Granick, S. (1966) J. Biol. Chem., 241, 1359.

65. Strik, J.J.T.W.A. (1973) Enzyme, 16, 211.

66. Vos, J.G., Maas, H.L. van der Musch, A. and Ram, E. (1971) Toxicol. Appl. Pharmacol., 18, 944.

67. Strik, J.J.T.W.A. (1978) Environ. Hlth. Perspect., 23, 167.

68. Vos, J.G., Strik, J.J.T.W.A., Holsteijn, C.W.M. and Pennings, J.H. (1971) Toxicol. Appl. Pharmacol., 20, 232.

69. Goldstein, J.A., Friesen, M., Scotti, T.M., Hickman, P., Hass, J.R. and Bergman, H. (1978) Toxicol. Appl. Pharmacol., 46, 633.

70. Sinclair, P.R. and Granick, S. (1974) Biochem. Biophys. Res. Commun., 61, 124.

71. Lissner, R., Goerz, G., Eichenauer, M.G. and Ippen, H. (1975) Biochem. Pharmacol., 24, 1729.

72. Farber, T.M. and Baker, A. (1974) Toxicol. Appl. Pharmacol., 29, 102.

73. Alvares, A.P., Fischbein, A., Anderson, M.D. and Kappas, A. (1977) Clin. Pharmacol. Ther., 22, 140.

74. Pużyńska, L., Debets, F.M.H. and Strik, J.J.T.W.A., this issue.

75. Goldstein, J.A., Hickman, P., Bergman, H., McKinney, J.D. and Walker, M.P. (1977) Chem.-Biol. Interactions, 17, 69.

76. Dent, J.G. (1978) Environ. Hlth. Perspect., 23, 301.

77. Babish, J.G. and Stoewsand, G.S. (1977) J. Toxicol. Environ. Hlth., 3, 673.

78. Greig, J.B. (1972) Biochem. Pharmacol., 21, 3196.

79. Greig, J.B. and De Matteis, F. (1973) Environ.Hlth. Perspect., 5, 211.

80. Lucier, G.W., McDaniel, O.S., Hook, G.E.R., Fowler, B.A., Sonawane, B.R. and Faeder, E. (1973) Environ. Hlth. Perspect., 5, 199.

81. Jones, K.G. and Sweeney, G.D. (1977) Res. Comm. Chem. Pathol. Pharmacol., 17, 631.

82. Kerklaan, P.R.M., Strik, J.J.T.W.A. and Koeman, J.H., this volume.

83. Mehendale, H.W., Fields, M. and Matthews, H.B. (1975) J. Agric. Food Chem., 23, 261.

84. Lui, H. and Sweeney, G.D. (1975) FEBS Letters, 51, 225.

85. Engst, R., Macholz, R.M. and Kujawa, M. (1976) Bull. Environ. Contam. Toxicol., 16, 248.

86. Renner, G. and Schuster, K.P. (1977) Toxicol. Appl. Pharmacol., 39, 355.

87. Koss, G., Strik, J.J.T.W.A. and Kan, C.A. (1978) in Industrial and Environmental Xenobiotics: in vitro versus in vivo biotransformation and toxicity, Fouts, J.R. and Gut, I. (eds.), Excerpta Medica, Amsterdam, p. 211.

88. Sanborn, J.R., Childers, W.F. and Hansen, L.G. (1977) J. Agr. Food Chem., 25, 551.

89. Yang, R.S.H., Coulston, F. and Golberg, L. (1975) J. of the AOAC, 58, 1197.

90. Rozman, K., Mueller, W. Iatropoulos, M., Coulston, F. and Korte, F. (1975) Chemosphere, 5, 289.

91. Kohli, J., Jones, D. and Safe, S. (1976) Can. J. Biochem., 54, 203.

92. Iatropoulos, M.J. Milling, A., Müller, W.F., Nohynek, G. and Rozman, K. (1975) Environ. Res., 10, 384.

93. Villanueva, E.C., Jennings, R.W., Burse, V.W. and Kimbrough, R.D. (1974) J. Agr. Food Chem., 22, 916.

94. Lui, H., Sampson, R. and Sweeney, G.D. (1976) in Porphyrins in Human Diseases, Doss, M. (ed.), S. Karger AG, Basel, p. 405.

95. Koss, G. and Koransky, W. (1978) in Pentachlorophenol, Ranga Rao, K.; (ed.) Plenum Press, New York, p. 131.

96. Parke, D.V. (1968) The Biochemistry of Foreign Compounds, Pergamon Press, Oxford.

97. Kószó, F., Siklósi, Cs. and Simon, N. (1978) Biochem.Biophys. Res.Commun.,, 80, 781.

98. Arrhenius, E., Renberg, L., Johansson, L. and Zetterqvist, M. (1977) Chem.-Biol. Interactions, 18, 35.

99. Tulp, M.Th.M., Sundstrom, G. and Hutzinger, O. (1976) Chemosphere, 5, 425.

100. Safe, S., Jones, D. and Hutzinger, O. (1976) J. Chem. Soc. Perkin trans., 1, 357.

101. Weinbach, E.C. (1954) J. Biol. Chem., 210, 545.

102. Loomis, W.S. (1949) Fed. Proc., 8, 220.

103. Jacobson, I. and Yllner, S. (1971) Acta Pharmacol., 29, 513.

104. Carlson, G.P. (1978) Toxicology, 11, 145.

105. Arrhenius, E. (1974) in Chemical Carcinogenesis Essays, Montesano, R. and Tomatis, L. (eds.), IARC Scientif. Publ. no. 10, Lyon, p. 25.

106. Arrhenius, E. (1969/70) Chem.-Biol. Interactions, 1, 381.

107. Maines, M.D. (1977) in Microsomes and Drug Oxidations, Ullrich, V. (ed.) Pergamon Press, Oxford, p. 543.

108. Dam, K. van, and Meyer, A.J. (1971) Ann. Rev. Biochem., 14, 115.

109. Lüllmann, H., Lüllmann-Rauch, R. and Reil, G.H. (1973) Virchows Arch., B12, 91.

110. Seydel, J.K. and Wassermann, O. (1973) Naunyn-Schmiedeberg's Arch. Pharmacol., 279, 207.

111. Hruban, Z. (1976) Environ. Hlth. Perspect., 16, 111.

112. Arrhenius, E., Renberg, L. and Johansson, L. (1977) Chem.-Biol. Interactions, 18, 23.

113. Chasseaud, L.F. (1976) in Glutathione, metabolism and function, Arias, I.M. and Jacoby, W.B. (eds.), Raven Press, New York, p. 77.

114. Larsen, G.L. and Bakke, J.E. (1978) Abstract presented at IUPAC Pesticide Congress, Zürich, July 1978.

115. Rimington, C. and Ziegler, G. (1963) Biochem. Pharmacol., 12, 1387.

116. Wolf, M.A., Lester, R. and Schmid, R. (1962) Biochem. Biophys. Res. Comm., 8, 278.

117. Recknagel, R.O. and Glende, E.A. (1973) CRC Crit. Rev. Toxicol., 2, 263.

118. Metcalfe, R.L., Kapoor, I.P., Lu, P.Y., Schuth, C.K. and Sherman, P. (1973) Environ. Hlth. Perspec., 4, 35.

119. Tulp, M.Th.M., Bruggeman, W.A. and Hutzinger, O. (1977) Experientia, 33, 1134.

120. Ahlborg, U.G. and Thunberg, T. (1978) Arch. Toxicol., 40, 55.

121. Daly, J.W., Jerina, D.M. and Witkop, B. (1972) Experientia, 28, 1129.

122. Mizutani, T., Mio, T. and Sumino, K. (1975) in Proc. 2nd Symposium on Environmental Toxicology, Gifu, Japan, 81.

123. Jensen, S. and Jansson, B. (1976) Ambio, 5-6, 257.

124. Koss, G., Seubert, S., Seubert, A., Koransky, W. and Ippen, H. (1977) Abstract Joint Meeting German and Italian Pharmacologists, Venice, Oct 1977.

125. Ahlborg, U.G., Lindgren, J.E. and Mercier, M. (1974) Arch. Toxicol., 32, 271.

126. Oesch, F. and Daly, J. (1972) Biochem. Biophys. Res. Commun., 46, 1713.

127. Oesch, F. (1976) J. Biol. Chem., 251, 79.

128. Dent, J.G. et al. (1977) Life Sci., 20, 2075.

129. Dent, J.G., McCormack, K.M., Rickert, D.E., Cagen, S.Z., Melrose, P. and Gibson, J.E. (1978) Toxicol. Appl. Pharmacol., 46, 727.

130. McCormack, K.M., Kluwe, W.M., Sanger, V.L. and Hook, J.B. (1978) Environ. Hlth. Perspect., 23, 153.

131. San Martin de Viale, L.C., Viale, A.A., Nacht, S. and Grinstein, M. (1978) Clin. Chem. Acta, 28, 13.

132. Elder, G.H., Evans, J.O. and Matlin, S. (1976) in Porphyrins in Human Diseases, Doss, M. (ed.), Karger, Basel, p. 424.

133. Strik, J.J.T.W.A. (1973) Thesis, Utrecht, The Netherlands.

134. De Matteis, F., Prior, B.E., Rimington, C. (1961) Nature (Lond.), 191, 363.

135. San Martin de Viale, L.C., Tomio, J.M., Ferramola, A.M., Sancovich, H.A. and Tigier, H.A. (1972) Nat. Meeting Arg. Soc. Biochem. Invest., Abstr. 8.

136. San Martin de Viale, L.C., Rios de Molina, M. del C., Wainstock de Calmanovici, R. and Tomio, J.M. (1976) in Porphyrins in Human Diseases, Doss, M. (ed.), Karger, Basel, p. 445.

137. Elder, G.H., Evans, J.O. and Matlin, S. (1976) Clin. Sci. Molec. Med., 51, 71.

138. Blekkenhorst, G.H., Pimstone, N.R., Webber, B.L. and Eales, L. (1976) Ann. Clin. Res., 8, suppl. 17, 108.

139. Louw, M., Neethling, A.C., Percy, V.A., Carstens, M. and Shanley, B.C. (1977) Clin. Sci. Molec. Med., 53, 111.

140. Romeo, G. and Levin, E.Y. (1971) Biochim. Biophys. Acta, 230, 330.

141. Simon, N., Siklosi, C. and Koszo, F. (1976) in Porphyrins in Human Diseases, Doss, M. (ed.), Karger, Basel, p. 432.

142. Simon, N., Dobozy, A. and Berko, G. (1970) Arch. Klin. Exp. Derm., 238, 38.

143. Vos, J.G. and Notenboom-Ram, E. (1972) Toxicol. Appl. Pharmacol., 23, 563.

144. Hacking, M.A., Budd, J. and Hodson, K. (1978) Can. J. Zool., 56, 477.

145. Corbett, T.H., Simmons, J.L., Kawanishi, H. and Endres, J.L. (1978) Environ. Hlth. Perspect., 23, 275.

146. Norback, D.H. and Allen, J.R. (1969) Lab. Invest., 20, 338.

147. Stenger, R.J. (1966) J. Ultrastruct. Res., 14, 240.

148. Herdson, P.B., Garvin, P.J. and Jennings, R.B. (1964) Lab. Invest., 13, 1032.

149. Ortega, P. (1966) Lab. Invest., 15, 657.

150. Hutterer, F., Klion, F.M., Wengraf, A., Schaffner, F. and Popper, H. (1969) Lab. Invest., 20, 455.

151. Thys, O., Hildebrand, J., Gerin, Y. and Jaques, P.J. (1973) Lab. Invest., 28, 70.

152. Verschuuren, H.G. (1967) Fd. Cosmet. Toxicol., 5, 450.

153. Elder, G.H. (1978) in Heme and Hemoproteins, De Matteis, F. and Aldridge, W.N. (eds.), Springer-Verlag, Berlin, p. 157.

154. Vos, J.G., Botterweg, P.F., Strik, J.J.T.W.A., Koeman, J.H. (1972) TNO-nieuws, 27, 599.

155. Förlin, L. and Strik, A. (1978) in Industrial and Environmental Xenobiotics: in vitro versus in vivo Biotransformation and Toxicity, Fouts, J.R. and Gut, I. (eds.), Excerpta Medica, Amsterdam, p. 302.

156. Carlson, G.P. (1977) Experientia, 33, 1627.

157. Doss, M. (1968) FEBS Letters, 2, 127.

158. Doss, M. (1969) Z. klin. Chem. u. klin. Biochem., 7, 133.

159. Sassa, S. and Granick, S. (1970) Proc. Nat. Acad. Sci. USA, 67, 517.

160. Granick, S., Sinclair, P., Sassa, S. and Grieninger, G. (1975) J. Biol. Chem., 250, 9215.

161. Roomi, M.W. (1975) J. Med. Chem., 18, 457.

162. Kawanishi, S., Mizutani, T. and Sano, S. (1978) Biochim. Biophys. Acta, 540, 83.

163. Grisham, J.W., Charlton, R.K. and Kaufman, D.G. (1978) Environ. Hlth. Perspect., 25, 161.

164. Ippen, H. Aust, D. and Goerz, G. (1972) Arch. Derm. Forsch., 245, 305.

165. Grant, D.L., Shields, J.B. and Villeneuve, D.C. (1975) Bull. Environ. Contam. Toxicol., 14, 422.

166. San Martin de Viale, L.C., Tomio, J.M., Ferramola, A.M., Sancovich, H.A. and Tigier, H.A. (1976) in Porphyrins in Human Diseases, Doss, M. (ed.), Karger, Basel, p. 453.

PROCEDURES FOR PORPHYRIN-, THIOETHER ASSESSMENT AND DIAGNOSIS OF CHRONIC HEPATIC PORPHYRIA

Chemical Porphyria in Man, J.J.T.W.A. Strik and J.H. Koeman eds.

TOTAL PORPHYRIN ANALYSIS IN URINE

E.G.M. HARMSEN and J.J.T.W.A. STRIK
Department of Toxicology, Agricultural University, Wageningen, The Netherlands.

Sample handling

Until analysis the urines are stored at -50 °C. Before defreezing a total check of porphyrins can be made by placing the still frozen urines under U.V. light (Desaga 366 nm long range intensive U.V. source). The porphyrins appear red or orange under U.V. light and in the frozen urines, are mostly sedimented to the bottom of the flask.

After defreezing, the urine samples have to be shaken vigorously to resuspend the sediment.

Isolation of porphyrins

Select chromatography columns* (20 cm x 12 mm outside Ø) with stopcocks. Plug with cotton and add 1.5 g anion-exchange resin (AG I - X8, 50-100 mesh chloride form, Bio.Rad. Laboratories, Richmond, Calif.).

Wet columns with distilled water and leave overnight. Drain the colums. Pipet 1 ml urine into the upper column wall, directly above the resin. Wash the resin with 8 ml distilled water to remove interfering substances. Place culture tubes under the columns.

Porphyrins are eluted 2 times in 2 ml fractions of 3N hydrochloric acid. The addition of the second portion of HCl is made after the first has run through. Clean the columns by flushing with about 30 ml distilled water. The wash water must be pH 5-6 before the columns are ready for re-use.

These columns can be re-used many times. However, the first run is required to adapt the resin. Discard the HCl fractions and start again. Keep the taps of the columns open during HCl eluting in order to avoid light exposure as much as possible, as this might significantly decrease the quantity of the isolated porphyrins[1].

Measurement and calculation

Measure the absorbance of the eluate (4 ml 3N HCl) in a U.V. spectrophotometer (Beckman Acta CIII double beam spectrophotometer) at 430, 380 and 403 nm.

*brown glass is advisable, light may destroy the porphyrins.

Correct the porphyrin absorbance as follows:

$$A_s \text{ (corrected) } \frac{2\,A_{403} - (A_{430} + A_{380})}{k}$$

Correctional factor k = 1.8. This factor corrects the absorbance values of free porphyrin acids from biological material. Millimolair extinction coefficients in 3N HCl:

uroporphyrin 507 $mmol^{-1}\ cm^{-1}$

coproporphyrin 428 $mmol^{-1}\ cm^{-1}$

A mean molecular weight of 691 and a mean millimolar extinction coefficient of 444 are determined for uro- plus coproporphyrin by taking the average ratio of uro- to coproporphyrin of about 1 to 4. These values are used in the following calculation of total porphyrins:

$$\frac{A_s \text{ (corrected)} \times 691 \times \text{HCL volume (ml)}}{444 \times d\text{ (cm)} \times \text{urine sample volume (ml)}}$$

Recovery of total porphyrins

To test the recovery of porphyrins from the ion-exchange columns, spike about 10 urine samples with a known concentration coproporphyrin - I (Sigma chemical company) and analyse them again as described above. Correct the absorbances and calculate the total porphyrin.

The total porphyrin in the standard can be calculated as follows

$$\frac{A_s \text{ (corrected)} \times 654 \times \text{HCl volume (ml)}}{428 \times d\text{ (cm)} \times \text{volume of standard (ml)}}$$

The recovery can be calculated as follows:

$$\text{Recovery} = \frac{\text{total porphyrin (spiked urine)} - \text{total porphyrin (urine)}}{\text{total porphyrin (standard)}}$$

REFERENCE

Doss, M. (1974) Porphyrins and Porphyrin Precursors, in M.C. Curtius and M. Roth (eds.), Clinical Biochemistry, Principles and Methods, De Gruyter, Berlin, Vol. II, p. 1339.

Chemical Porphyria in Man, J.J.T.W.A. Strik and J.H. Koeman eds.

URINARY PORPHYRIN PATTERN ANALYSIS

J.J.T.W.A. STRIK and E.G.M. HARMSEN
Department of Toxicology, Agricultural University, Wageningen, The Netherlands

Extraction of the porphyrin methyl-esters

Preparation for analysis

Place 20 ml of urine (acidified to pH 4 with glacial acetic acid) in a 50 ml centrifuge tube with 0,8 g talcum (pharmaceutical grade). Close the tube with parafilm and shake vigorously for 30 s to allow the adsorption of the porphyrins onto the talc. Sediment the talc (3200 RPM, 3 min) and discard the supernatant. Add 5 ml methanol to the pellet and stir with a glass rod. Shake the suspension and centrifuge as above. Decant the methanol supernatant into a second centrifuge tube which contains 0,4 g talcum. The pellet in the first tube is then treated with another 5 ml methanol, and after the centrifugation, this methanol is also added to the second tube. Then the second centrifuge tube is shaken for about 20 s and centrifuged. The methanol supernatant is discarded. Both samples can be esterified.

Esterification

Both the talc layers are wetted with about 4 ml methanol sulfuric acid (95 ml methanol + 5 ml sulfuric acid). Close the tubes very well with parafilm and place them in the dark for 40 min at 37 ^{o}C, or 8 hours at room temperature.

Extraction

Add 3 ml chloroform to the esterification mixture, and resuspend the talcum layer with a glass rod. The mixture is centrifuged and the supernatant is decanted into a 100 ml separating funnel. The talcum layer is then extracted two more times with 3 ml chloroform. The chloroform supernatants are collected in the separating funnel, which is then filled to about 80 % of its volume with distilled water. It is closed and shaken vigorously. After the phases have separated, the chloroform phase is with-drawn into culture tubes. The aqueous phase in the separatory funnel is extracted twice more with 2 ml portions of chloroform. The chloroform phases are combined in the culture tubes and shaken carefully with about 3 ml saturated $NaHCO_3$ (to increase pH). The upper aqueous phase is discarded, and the chloroform extracts are washed twice more with about 4 ml distilled water. The aqueous phase (after second washing) must be neutral (check with pH paper).

The chloroform extract is dried with 2 to 3 spatula tips of dry sodium sulfate, and filtered through a folded paper filter. The chloroform containing the porphyrins is collected into small tubes.

At this point the dissolved porphyrins can be observed by illuminating the tubes in a dark room with a UV lamp (Desaga 366 nm long range intensive source). The chloroform is evaporated under N_2 at room temperature.

Thin Layer Chromatography

Application of the porphyrin methyl esters: The dried porphyrin methyl esters are dissolved in a small volume (50 - 100 µl) chloroform. Observe here also the macroscopic fluorescence, spot the aluminum backed porphyrin solutions in small lines onto silica gel plates (37361, DC-karten SI, 20 x 20 cm, Riedel-De Haën Aktiengesellschaft, Seelze-Hannover).

Development of the chromatograms: 3 solvent systems are used (renew daily).

1. Petroleum ether (40-60)-diethylether (3/1, v/v). Develop so that the solvent front moves to the top of the sheet. This process removes most of the lipids, which migrate with the solvent front.
2. Chloroform-methanol (130/20, v/v). Develop only 1 cm to form a new start line. The chromatogram must be dry: 40 min. at room temperature, in dark, or 12 min under a cold-air hair dryer.
3. Benzene-ethylactetate-methanol (85/13.5/1.5, v/v). This solvent system separates the porphyrins. Develop so that the solvent front has moved 2/3 of the way to the top.

Location of the porphyrins: Direct after developing in the last solvent the TLC plates have to be inspected visually under U.V. light (Desaga 366 nm) and the relative amounts of the porphyrins present have to be recorded (lined by pencil).

Quantification of the spots: Scrape off the lined spots of the different porphyrins on the TLC plates. Dissolve the scrapings in chloroform and centrifugate. Measure the supernatant in a double beam spectrophotometer (Beckman Acta CIII) between 380 and 430 nm. From these spectra the different porphyrin % concentrations can be calculated.

REFERENCE

Doss, M. (1974). Porphyrins and Porphyrin Precursors, in: M.C. Curtius and M. Roth (eds.), Clinical Biochemistry, Principles and Methods, De Gruyter, Berlin, Vol. II, p. 1339.

Chemical Porphyria in Man, J.J.T.W.A. Strik and J.H. Koeman eds.

QUALITY CONTROL STUDIES ON URINARY δ-AMINOLEVULINIC ACID, PORPHOBILINOGEN AND PORPHYRIN ANALYSIS

R. v. TIEPERMANN and M. DOSS
Department of Clinical Biochemistry, Faculty of Medicine of the Philipp University, Marburg, F.R.G.

SUMMARY

Data are presented on quality control studies of urinary δ-aminolevulinic acid, porphobilinogen and porphyrin determinations. Porphyrin precursors were separated by ion exchange chromatography, and porphyrins by thin-layer chromatography; all the metabolites were quantitated by recording spectrophotometry. In series, over several weeks the variability coefficient of δ-aminolevulinic acid lay between 5 and 7 %, of porphobilinogen between 6 and 8 % and uro- and coproporphyrin between 4 and 6 % ('precision from day to day', n between 9 and 21). For quality control native urine samples (pH 5.5 - 6.5) were frozen at -30 ^{o}C: under these conditions no decrease of the concentrations of ALA was observed up to 30 weeks, of PBG up to 10 weeks or of porphyrins up to 16 weeks. It can be concluded that porphyrin precursors and porphyrins are stable within these time intervals.

INTRODUCTION

Not only routine tests in Clinical Chemistry, but also special methods in Clinical Biochemistry should be judged according to the concepts of quality control elaborated by the 'International Federation of Clinical Chemistry'[1]. The duties of porphyria centers for research, diagnosis and treatment[2] include also the improvement of accuracy of laboratory measurements of the metabolites of pyrrol, porphyrin and heme biosynthesis as well as of their enzymes.

During the 'International Symposium on Clinical Biochemistry on Diagnosis and Therapy of Porphyrias and Lead Intoxication' (on the occasion of the 450-year celebrations of the Philipp University of Marburg in 1977), a session was held on 'Quality Control in Porphyrin Laboratories' - as suggested and promoted by the 'German Association of Clinical Chemistry'.

The results are presented:

a. Introduction of internationally accepted quality control in the clinical biochemistry of porphyrins and porphyrin precursors[3].
b. Recommendations for porphyrin laboratories and clinical chemists in porphyria research[4].

During the discussion it was agreed that the results on quality control checks developed and performed in the Department of Clinical Biochemistry in Marburg should be published in detail[4]: in the course of development and establishment of new principles and methods for both clinical and experimental studies, we have since 1968, considered and elaborated standardization, precision, specifity, recovery studies and long-range stability of laboratory performance in the field of a porphyria center[5-9]. Part of the details most interesting for clinical biochemical diagnostics are presented here.

METHODS AND RESULTS

Urine samples for porphyrin precursors and porphyrin analysis were prepared according to the rules used in our laboratory for ten years which agree with the recommendations discussed at international level (published in J. Clin. Chem. Clin. Biochem.[4]). Samples (at pH 5.5 - 6.5) for longterm quality control were frozen at -30 °C. Porphyrin precursors were separated by ion exchange chromatography[5] and porphyrins by thin-layer chromatography[6,7]; both were analyzed quantitatively by recording spectrophotometry[5,8].

Results of quality control studies on the determination of urinary ALA and PBG are compiled in table 1 and those of uro- and coproporphyrin in table 2. The data represent the variability coefficients for the 'precision from day to day' in various series at different times in 1975, 1976 and 1977. The results of longterm studies including simultaneous stability tests of the metabolites ALA and PBG are illustrated in fig. 1. When comparing long-term studies for ALA and PBG over 70 weeks, one notices (fig. 1a and 1b), that ALA undergoes only a small fall in concentration. Therefore the regression line for the PBG analyses shows a negative increase of 0.1, but for ALA, however, it is only 0.01. Because of PBG's lesser stability the variability coefficient of PBG is always higher than that of ALA. For the thin-layer chromatographic porphyrin determinations the variability coefficient lies below 5 % (table 2). The samples frozen at -30 °C showed no decrease in concentration within 17 weeks (fig. 2).

DISCUSSION

The variability coefficient for ALA and PBG lay between 5 and 7 % in a series of 13-20 determinations between 10 and 20 weeks. This result is not yet optimal, however, considering the difficulty of the method, it is acceptable and lies within the range demanded for special biochemical analysis. The data reported here are actually results of 'precision from week to week', because we performed

TABLE 1

RESULTS OF QUALITY CONTROL STUDIES ON THE DETERMINATION OF URINARY δ-AMINOLEVULINIC ACID (ALA) AND PORPHOBILINOGEN (PBG) BY ION EXCHANGE CHROMATOGRAPHY AND SPECTROPHOTOMETRY[5]: 'PRECISION FROM DAY TO DAY' IN DIFFERENT SERIES

Series	Substance	μmol/l ($\bar{x} \pm s$)	n	VC* (%)
A	ALA	112 ± 7	20	6.0
	PBG	62 ± 5	20	7.8
B	ALA	113 ± 6	21	5.4
	PBG	54 ± 4	21	7.4
C	ALA	114 ± 8	13	6.7
	PBG	39 ± 3	13	7.8

*Variability coefficient

TABLE 2

RESULTS OF QUALITY CONTROL STUDIES ON THE DETERMINATION OF URO- AND COPROPORPHYRIN BY TLC AND SPECTROPHOTOMETRY[6]: 'PRECISION FROM DAY TO DAY' IN SERIES OF ABOUT FOUR WEEKS EACH

Series	Substance	nmol/l ($\bar{x} \pm s$)	n	VC* (%)
A	URO	162 ± 8	10	4.9
	COPRO	395 ± 20	10	5.0
B	URO	171 ± 8	18	4.6
	COPRO	399 ± 17	18	4.4
C	URO	167 ± 7	11	4.2
	COPRO	391 ± 24	11	6.2
D	URO	168 ± 10	9	5.7
	COPRO	391 ± 14	9	3.5
E	URO	166 ± 8	10	5.0
	COPRO	366 ± 17	10	4.6

*Variability coefficient

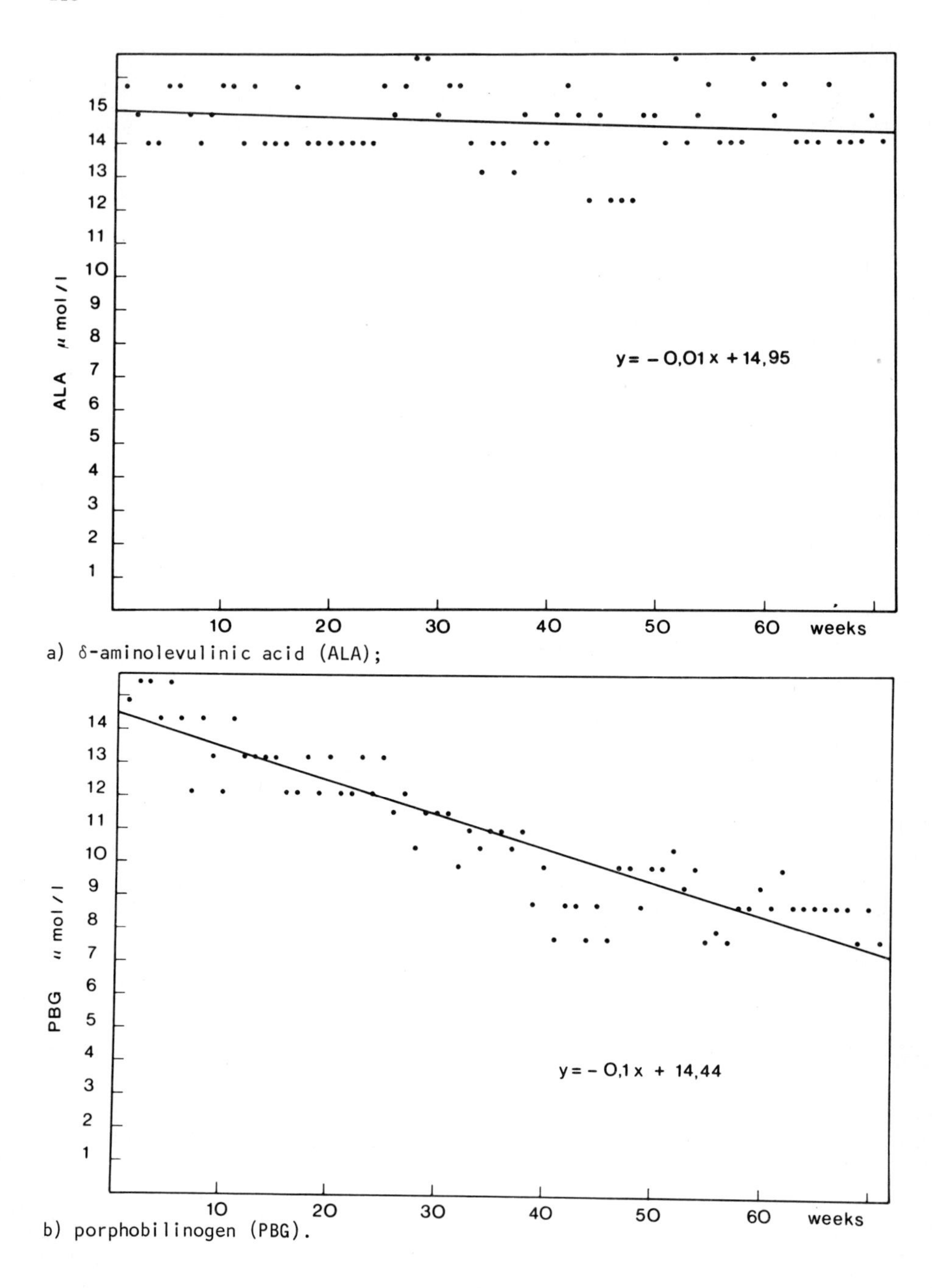

a) δ-aminolevulinic acid (ALA);

b) porphobilinogen (PBG).

Fig. 1. Quality control studies on urinary porphyrin precursors analyzed by ion exchange chromatography and spectrophotometry[5].

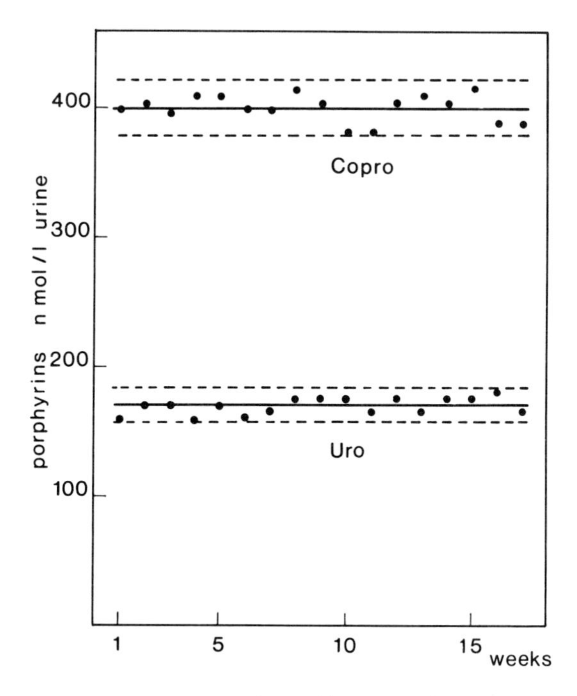

Fig. 2. Quality control studies on urinary uro- and coproporphyrin analyzed by TLC and spectrophotometry[6-8].
—— $\bar{x}$, ---- 2s range.

the tests for porphyrin precursors once a week. In a real series of 'precision from day to day' one obtains a variability coefficient between 2.6 and 3.5 (n=10)[5]. Because of the different stability of the two porphyrin precursors, one could probably optimize the results of quality control by lyophilizing the samples. The quality control data for uro- and coproporphyrin are acceptable and lie within the same range earlier obtained for thin-layer chromatographic analysis in combination with fluorometric quantitative measurement[7]. The kinetics of the destruction of the individual porphyrins depends on the number of carboxylic acid side chains, as observed in fluorometric studies in situ[9]. An exponential rise in stability of the porphyrins was found as the number of side chains increased[9]. These results are valid for exposure of porphyrins to light and heat. From this work the conclusion has been drawn that porphyrin analysis should not be performed in daylight or other direct light. Because light is a main factor for destroying porphyrins, including neon light[9], we are very anxious to avoid light exposure during porphyrin analysis and therefore prefer working in a darkened laboratory. In general the ten-year quality control study of the most important metabolites of porphyrin synthesis in urine (ALA, PBG, uro- and copro-porphyrin) shows that relatively complex and complicated manual clinical-biochemical methods may give accurate results under conditions of good laboratory practice.

In future quality control work has to be extended to fecal and erythrocyte porphyrins as well as to certain enzymes involved in the porphyrin pathway. For instance data from uroporphyrinogen synthase activity calculation in hemolysates are available and seem to be acceptable[10]. Similar studies on uroporphyrinogen decarboxylase are in progress.

ACKNOWLEDGEMENT

The experimental work of these studies was supported by the Deutsche Forschungsgemeinschaft, Bonn, Bad Godesberg.

REFERENCES

1. The International Federation of Clinical Chemistry, Committee on Standards (1976) and (1977) Recent lists of references are included in: Clin. Chem. 22, 1922 and J. Clin. Chem. Clin. Biochem., 15, 95.
2. Doss, M. (1978) J. Clin. Chem. Clin. Biochem., 16, 77.
3. Doss, M. and Schwartz, S. (1978) J. Clin. Chem. Clin. Biochem., 16, 76.
4. Doss, M. and Schwartz, S. (1978) J. Clin. Chem. Clin. Biochem., 16, 76.
5. Doss, M. and Schmidt, A. (1971) Z. klin. Chem. klin. Biochem., 9, 99.
6. Doss, M. (1970) Z. klin. Chem. klin. Biochem., 8, 197.
7. Doss, M. (1970) Z. Anal. Chem., 252, 104.
8. Doss, M. (1971) Anal. Biochem., 39, 7.
9. Doss, M. and Ulshöfer, B. (1971) Biochim. Biophys. Acta, 237, 356.
10. Doss, M. and Tiepermann, R. v. (1978) J. Clin. Chem. Clin. Biochem., 16, 1.

Chemical Porphyria in Man, J.J.T.W.A. Strik and J.H. Koeman eds.

NORMAL RANGES OF PORPHYRINS AND PRECURSORS IN HUMAN TISSUE, URINE AND FECES

M. DOSS

Department of Clinical Biochemistry, Faculty of Medicine of the Philipp University, Marburg, F.R.G.

Urine (μg/24 hr Urinevolume)	
δ-Aminolevulunic acid	250 - 6400
Porphobilinogen	100 - 1700
Uroporphyrin	3 - 24
Heptacarboxylicporphyrin	0 - 3
Hexacarboxylicporphyrin	0 - 2
Pentacarboxylicporphyrin	0 - 4
Coproporphyrin	14 - 78
Tricarboxylicporphyrin	0 - 1
Dicarboxylicporphyrin	0 - 1
Coproporphyrin-Isomer III	69 - 83 %
Coproporphyrin-Isomer I	17 - 31 %
Faeces (μg/g Dry Weight)	
X-Porphyrin	0 - 4
Uroporphyrin	1 - 6
Heptacarboxylicporphyrin	0 - 2
Hexacarboxylicporphyrin	0 - 1
Pentacarboxylicporphyrin	0 - 3
Isocoproporphyrin	0
Coproporphyrin	3 - 24
Tricarboxylicporphyrin	0 - 10
Protoporphyrin	12 - 85
Plasma (μg/100 ml)	
Uroporphyrin	0 - 0.1
Coproporphyrin	0 - 0.2
Protoporphyrin	0 - 0.8
Erythrocytes (μg/100 ml)	
Uroporphyrin	0 - 0.1
Coproporphyrin	0 - 2.0
Protoporphyrin	5 - 36.0
Liver (μg/g Dry Weight)	
Uroporphyrin	0 - 0.1
Coproporphyrin	0 - 0.3
Protoporphyrin	0.2 - 1.3

Chemical Porphyria in Man, J.J.T.W.A. Strik and J.H. Koeman eds.

CLASSIFICATION OF PORPHYRIAS AND SECONDARY COPROPORPHYRINURIAS

M. DOSS
Department of Clinical Biochemistry, Faculty of Medicine of the Philipp University, Marburg, F.R.G.

I. Erythropoietic Porphyrias

A. Porphyria congenita erythropoietica (autosomal recessive)
B. Protoporphyria erythro(hepato)poietica (autosomal dominant)

II. Hepatic Porphyrias

A. Porphyria acuta intermittens (autosomal dominant)
B. Coproporphyria hereditaria (autosomal dominant)
C. Porphyria variegata ("mixed hepatic porphyria"; autosomal dominant)
D. 1 Chronic hepatic Porphyrias including Porphyria cutanea tarda (non-hereditary or hereditary)

2 Paraneoplastic Porphyria cutanea tarda (e.g. liver- or prostatacarcinoma)

3 Symptomatic chronic hepatic Porphyrias (caused e.g. by Hexachlorobenzene, Vinylchloride)*

III. Secondary (symptomatic) Coproporphyrinurias, caused by:

A. Intoxications (e.g. alcohol, chemicals, polyhalogenated aromatics, heavy metals - esp. lead)
B. Liver diseases (e.g. cirrhosis, hepatitis, fatty liver, cholestasis, alcohol liver syndrome, hemochromatosis, drug induced liver damage)
C. Blood diseases (e.g. anemias: hemolytic, drug induced, sideroachrestic, sideroblastic, aplastic; ineffective erythropoiesis, leucemia, hemoblastosis)
D. Infectious diseases
E. Diabetes mellitus
F. Iron metabolism disturbances (e.g. hemosiderosis, hemochromatosis)
G. Inherent Hyperbilirubinemia (e.g. Dubin-Johnson-Syndrome, Rotor-Syndrome)
H. Malignant tumors
I. Heart attacks
J. Drug induced side effects (analgetics, sedatives, hypnotics, antibiotics, sulfonyl-ureum-derivatives, sex-hormones (oestrogens) and narcosis)
K. Fasting

* This volume

© 1979, Elsevier/North-Holland Biomedical Press
Chemical Porphyria in Man, J.J.T.W.A. Strik and J.H. Koeman eds.

THE ESTIMATION OF MERCAPTURIC ACIDS AND OTHER THIOETHERS IN URINE

F. SEUTTER-BERLAGE, H.L. VAN DORP, H.J.J. KOSSE, J.M.T. HOOG ANTINK and M.A.P. WAGENAARS-ZEGERS

Institute of Pharmacology, University of Nijmegen, Nijmegen, The Netherlands.

INTRODUCTION

Among the chemical contaminants to which man is occupationally and environmentally exposed, the compounds with potentially alkylating properties[1,2,3,4] are most hazardous. From animal experiments it is known that many of these substances are able to cause severe tissue lesions and even exert mutagenous or carcinogenous effects.

One of the protective mechanisms the organism employs against electrophilic alkylating products, is the inactivation by reaction with the endogenous nucleophilic glutathione, spontaneously or by means of glutathion S-transferases[5,6]. The glutathione conjugates are further metabolized by cleavage of the glutamate and glycine residues, followed by acetylation of the resultant free amino-group of the cysteinyl residue, to produce the final product, a mercapturic acid. The mercapturic acids, i.e. S-alkylated derivatives of N-acetylcysteine, are then excreted into the bile or urine.

It was already in 1879 that Bauman and Preuss[7] as well as Jaffe[8] discovered that after administration of bromobenzene to animals an urinary excretion of bromobenzene mercapturic acid was evident.

Recently, we had the idea to develop a simple practical procedure to detect an enhanced urinary thioether excretion as a biological parameter of exposure to potentially alkylating agents[9]. The aim of the present report is to incorporate more details and some improvements. The method was evaluated further, moreover eating and smoking habits have been considered.

MATERIALS AND METHODS

Chemicals

5,5'-Dithiobis(2-nitrobenzoic acid) (DTNB), glutathione and oxidized glutathione was obtained from Boehringer Mannheim GmbH, Mannheim, West Germany, sodium borohydride, ethylene-diamine tetraacetic acid (EDTA), metaphosphoric acid and N-acetylcysteine from E. Merck A.G., Darmstadt, West Germany. Sodium citrate was purchased from Baker Chemicals N.V., Deventer, The Netherlands. All other chemicals were analytical grade.

Samples

Samples of freshly voided urine were frozen immediately at -18° without any additions and transported under refrigeration to the analytical laboratory.

Deproteinization

Urine samples were deproteinized according to the method of Beutler et al.[10] in the following modification: 2 ml urine samples were added to 3 ml of a reagent consisting of 120 g NaCl, 6.68 g metaphosphoric acid and 0.8 g of EDTA dissolved to 400 ml in distilled water. The reaction mixture was left 20 min. at 0° and then centrifuged at 3000 r.p.m. The analysis was performed with the supernatant.

Determination of sulfhydryl groups

The free SH-concentration in the solution was determined according to Ellman[11].

Free SH-groups: 0.25 ml of supernatant was mixed with 2 ml Sörensen phosphate buffer 0.5M pH 7.1 and 0.3 ml reagent solution, consisting of 40 mg DTNB and 1.12 g sodium citrate (2 H_2O) in distilled water to 100 ml (this solution could be used up to 13 weeks when stored at 4°C). The extinction was read at 412 nm against distilled water. A blank for the urine was measured by substitution of 0.3 ml distilled water for the reagent, a blank for the reagent by substitution of 0.25 ml distilled water for the urine supernatant.

The number of mMol SH eq/l was calculated as:

$$\frac{5}{2} \times \frac{2.55}{0.25} \times \frac{E_{det.} - (E_{bl.urine} + E_{bl.\ reagent})}{\text{mol. ex. coeff.}}$$

SH-groups after alkaline hydrolysis of -S- and -S-S-:
2 ml of supernatant was added to 0.5 ml 5M NaOH in a dark glass tube with screw cap which was closed under nitrogen and heated at 100° for 50 min.

The reaction mixture was cooled in ice during 10 min. and neutralized with 0.25 ml HCl 10M. The measurement was performed after 25 min. at 412 nm against distilled water with a mixture of 0.25 ml of the latter solution, 2 ml of the buffer and 0.3 ml of the reagent mentioned above. Again a blank for the urine was obtained by substitution of 0.3 ml distilled water for the reagent, a blank for the reagent by substitution of 0.25 ml distilled water for the hydrolyzed urine supernatant.

The number of total mMol SH eq./l was calculated as:

$$\frac{5}{2} \times \frac{2.75}{2} \times \frac{2.55}{0.25} \times \frac{E_{det.} - (E_{bl.\ urine} + E_{bl.\ reagent})}{\text{mol. ex. coeff.}}$$

The difference between the latter and the former calculated amount is the number of mMol SH eq/l set free from thioethers and disulphides.

The molar extinction coefficient, used for the calculations, was estimated by a series of standards (such as glutathione or N-acetylcysteine) with the same equipment and reagents as we used for urine analyses.

Creatinine assay

The amounts of SH-equivalents were related to the creatinine content of the urine. The creatinine concentration of each sample was assayed as described by Gorter and De Graaff[12].

RESULTS AND DISCUSSION

Precipitation of proteins

Deproteinization is unavoidable in pathologic urines or animal urines that have been contaminated with food. Under normal conditions urine does not contain polypeptides, however, it does contain oligopeptides[13,14] which are, like the mercapturic acids, not precipitated in a strongly acid medium. These oligopeptides, hence, contribute to the background readings.

Though trichloroacetic acid is the most usual precipitating agent, we chose metaphosphoric acid, as its use not only results in a less low pH but has the advantage to form part of the buffer in a later stage, thus preventing a unnecessary addition of chemicals.

Precipitating agents such as $Zn(OH)_2$ must be avoided as metals interfere with the colouring reagent[15] and are potential catalysts for the autoxidation of SH-groups. For this reason EDTA was added to the deproteinization reagent[16].

The colour reaction

Originally Ellman[17] used bis-(p-nitrophenyl) disulphide. Its insolubility in water was a reason to use DTNB instead, which is much better soluble[11]. The reaction is that of one half of the disulphide molecule with a RSH molecule to a mixed disulphide, resulting in the formation of 3-carboxy-4-nitrothiophenol, which in an alkaline medium gives the yellow coloured anion.

Heavy metals are able to form complexes with free SH-groups[16], which influences the formation of the colour. The colour depends also on the pH of the reaction mixture. According to Ellman[11] an optimum is reached at pH 8, however, at this pH the colour is not stable. In a weakly alkaline medium SH-groups are already easily autoxidized, the rate of this oxidation increases contineously with increasing pH. Some investigators have tried to overcome this difficulty by adding anti-oxidantia like Vitamin C[18,19]. However, vitamin C has a much higher redox-potential than glutathione[15], which means that dehydroascorbate oxidizes simple thiols to disulphides and is itself reduced to ascorbate. In the presence of ascorbate thiol autoxidation instead of being inhibited by EDTA may actually be stimulated because its complex with metal ions present can catalyze the conversion of ascorbate to dehydroascorbate[20,21,22,23] by means of air oxygen, the presence of which cannot be totally avoided.

From our control experiments we judged that it was better to keep the pH of the final reaction mixture as low as possible (pH 6.8 - 7.1) and not to add antioxidants.

The cleavage of the thioether bond in an alkaline medium

Substances such as R-S-R and R-S-S-R are decomposed by bases. The rate of fission of the thioether group of a mercapturic acid depends on the temperature and the nature of the substitutents. Aliphatic mercapturic acids are slower in reacting then those in which the sulphur atom is directly bond to an aromatic nucleus.

With a number of model substances a mean optimal time of fission appeared 50 min. at 100° in 1N NaOH. After a known exposure the circumstances of hydrolysis can be chosen optimally in accordance with those for synthetized and isolated mercapturic acids.

In the case of disulphides the exact mechanism of the alkaline cleavage is unknown. Two hypotheses have been proposed: (a) A nucleophilic attack by hydroxide ions on one of the sulphur atoms results in a hydrolytic cleavage to three equivalents of RS^- on one of RSO_2^-. (b) An abstraction of a proton by OH^- is followed by a β-elimination reaction (24 t/m 31). The formation of a persulphide group (RSSH) has been reported in connection with this mechanism.

For us this was one of the reasons to remove -S-S-bonds by reduction in a later study[37].

When the mercaptans resulting after hydrolysis are exposed to the air they can be either oxidized to sulphinic or sulphonic acids or give radical reactions[23]. The resulting substances do not react anymore with the Ellman reagent. For this reason the reaction is performed in dark vials under nitrogen. Nevertheless a too long hydrolysis time invariably results in losses. A table of recoveries for model-substances is shown in Table 1.

TABLE 1

RECOVERY FOUND AFTER TREATMENT OF SOME SYNTHETIZED MERCAPTURIC ACIDS UNDER THE CIRCUMSTANCES OF HYDROLYSIS (1N NaOH / 50 min. / 100°)

Mercapturic acid	Recovery %
bromobenzene-	89
benzene-	95
benzyl-	87
n-hexyl-	60
n-propyl-	42
iso-propyl-	43
allyl-	31

At the end of alkaline hydrolysis the reaction mixture is neutralized with HCl. In doing so volatile thiols can disappear from the reaction micture by evaporation (see Table 1). To avoid losses in cases where low boiling thiols are expected neutralisation should be performed in closed vials under refrigeration.

Background

To evaluate the background value of a normal thioether concentration in the urine 50 24h. samples of healthy volunteers (aged 18-27) without any medication or drug use and not exposed to an abnormal environmental pollution, were examined. We could not exclude a moderate use of cigarettes or alcohol. The mean value for the ratio μMol SH eq. / mMol creatinine was 54 ± 20 ($\bar{x} \pm$ S.D.) showing a considerable background value.

To obtain some information on the influence of exogenous factors in the appearance of high background values of a number of volunteers, from those mentioned above, one group smoking, the other non-smoking, were screened on urinary thioether excretion twice weekly during several weeks.

The results are represented in Table 2.

TABLE 2

EXCRETION OF µMOL SH eq. / mMOL CREATININE IN VOLUNTEERS

Smokers $\bar{x} \pm$ S.D.	n
57.0 ± 10.0	18
62.9 ± 6.9	16
52.4 ± 8.1	20
61.5 ± 9.6	7
Non-smokers	
42.1 ± 8.9	18
51.3 ± 7.5	12
48.6 ± 9.3	16
43.4 ± 10.4	6

A highly significant difference between smokers and non-smokers appeared here and in the evaluation of the results of urine analysis in a group of workers exposed to pesticides (see this volume).

Other exogenous influence must be investigated further. The occurrence of sulphur containing substances in food constituents, such as onions, cabbage species, pineapples[32,33,34] (chiefly disulphides and sulfoxides) and bean species[35,36] (S-alkylcysteine derivatives) is well documented.

In neutralization after hydrolysis in an open vial several of these normal contributions to the background value will partly or totally disappear. In the screening of a possible exposure to potentially alkylating substances that are metabolized to mercapturic acids, yielding volatile mercaptans, the contribution of food constituents in neutralization in a closed vial might be a serious disadvantage.

REFERENCES

1. Miller, J.A. (1970) Cancer Res., 30, 559.
2. Mitchell, J.R. and Jollow, D.J. (1973) Role of Metabolic Activation in Chemical Carcinogenesis and in Drug-induced Hepatic Injury. In Drugs and the Liver (Eds. W. Gerok and K. Sickinger), 3rd Int. Symp. Freiburg, Schattauer, Stuttgart, p. 395.
3. Ryser, H.J.P. (1971) New. Eng. J. Med., 285, 721.

4. Sims, P. and Grover, P.L. (1974) Epoxides in Polycyclic Aromatic Hydrocarbon Metabolism and Carcinogenesis. In: Advances in Cancer Research (Eds. G. Klein and S. Weinhouse). Vol. 20, Acad. Press., New York, p. 166.

5. Boyland, E. (1971) Mercapturic Acid Conjugation. In: Handbook of Exp. Pharmac. (Eds. B.B. Brodie and J.R. Gilette) Vol. 28, II, Springer, Berlin, p. 584.

6. Chasseaud, L.F. (1967) In: Glutathione: Metabolism and Function. (Eds. I.M. Arias and W.B. Jakoby), Raven Press., New York, p. 77.

7. Baumann, E. and Preusse, C. (1879) Ber., 12, 806.

8. Jaffe, M. (1879) Ber., 12, 1092.

9. Seutter-Berlage, F., Dorp, H.L. van, Kosse, H.G.J. and Henderson, P.Th. (1977) Int. Arch. Occup. Environ. Hlth., 39, 45.

10. Beutler, E., Duron, O. and Kelly, B.M. (1963) J. Lab. & Clin. Med., 61, 882.

11. Ellman, G.L. (1959) Arch. Biochem. and Biophys., 82, 70.

12. Gorter, E. and Graaff, W.C. de, In: Klinische Diagnostiek, Stenfert Kroese, Leiden, 7th ed., p. 440.

13. Documenta Geigy, Wissenschaftliche Tabellen. 6. Auflage, Ed. K. Diem, Basle, Switzerland.

14. Henning, N. (1966) Klinische Laboratoriumdiagnostik, III. Auflage Urban & Schwarzenberg, München-Berlin-Wien.

15. West, E.S., Todd, W.R., Mason, H.S. and Bruggen, J.T. van (1966) In: Textbook of Biochemistry, The MacMillan Company, New York, p. 912.

16. Martell, A.E. and Calvin, M. (1952) In: Chemistry of the metal chelate compounds. Englewood Cliffs, N.J., Prentice-Hall.

17. Ellman, G.L. (1958) Arch. Biochem. Biophys., 74, 443.

18. Baars, A.J., Dongen, E.W. van and Breimer, D.D. (1977) Pharmaceutisch Weekblad, 112, 1117.

19. Vainio, H., Savolainen, H. and Kilpikari, I. (1978) Brit. J. Ind. Med., 35, 232.

20. Pirie, A. and Heyningen, R. van (1954) Nature, 173, 873.

21. Grimble, R.F. and Hughes, R.E. (1968) Life Sciences, 7, 383.

22. Meacham, J. (1968) Exp., 24, 125.

23. Jocelyn, P.C. (1972) In: Biochemistry of the SH-group. New York, Acad. Press.

24. Danehy, J.P. and Hunter, W.E. (1967) J. Org. Chem., 32, 2047.

25. Gawron, O. and Odstrchel, G. (1967) J. Acad. Chem. Soc., 89, 3263.

26. Schneider, J.F. and Westley, J. (1969) J. Biol. Chem., 244, 5735.

27. Anderson, L.-O. and Berg, G. (1969) Biochim. Biophys. Acta, 192, 534.

28. Donovan, J.W. and White, T.M. (1971) Biochem., 10, 32.

29. Asquith, R.S. and Carthew, P. (1972) Biochim. Biophys. Acta, 278, 346.

30. Anderson, W.L. and Wetlaufer, D.B. (1975) Anal. Biochem., 67, 493.

31. Friderici, G., Dupré, S., Matarese, R.M., Solinas, S.P. and Cavallini, D. (1977) Int. J. Peptide Protein Res., 10, 185.

32. Hegnauer, R. Chemotaxonomie der Pflanzen (II). Birkhäuser Verlag, Basel, Schweiz.

33. Virtanen, A.I. and Matikkala, E.J. (1960) Hoppe-Seylers Z. Physiol. Chemie, 322, 8.

34. White, R.H. (1975) Science, 189, 810.

35. Zacharius, R.M., Morris, C.J. and Thompson, J.F. (1959) Arch. Biochem. Biophys., 80, 199.

36. Rinderknecht, H., Thomas, D. and Aslin, S. (1958) Helv. Chim. Acta, 41, 1.

37. Seutter-Berlage, F., Selten, G.C.M., Oostendorp, S.G.M.L. and Hoog Antink, J.M.T. The modified thioether test. This volume.

© 1979, Elsevier/North-Holland Biomedical Press
Chemical Porphyria in Man, J.J.T.W.A. Strik and J.H. Koeman eds.

THE MODIFIED THIOETHER TEST

F. SEUTTER-BERLAGE, G.C.M. SELTEN, S.G.M.L. OOSTENDORP and J.M.T. HOOG ANTINK
Institute of Pharmacology, University of Nijmegen, Nijmegen, The Netherlands

INTRODUCTION

The total amount of sulphur that is excreted in human urine is estimated at 1.32 g/24 h. The greater part of this excretion is formed by inorganic sulphate (1.17 g) and organic sulphates (0.09 g). A small part (65 $\pm$ 15) mg/24 h consists chiefly of mercaptans, disulphides, sulphur obtaining amino acids such as methionine and cysteine[1] and S-alkylcysteine derivatives (mercapturic acids)[2-12].

Among the latter are a few of endogenous origin, namely those originating from the conjugates of glutathione with dopa, oestrogens and prostaglandins [9,10,11,12].

The endogenous urinary mercapturic acids are normally present in a concentration that is far below the detection limit of our estimation. However, the endogenous urinary disulphides might interfere in our test in contributing considerably to the high background value[13].

We might be able to eliminate these disulphides by reduction with $NaBH_4$ before starting the analysis of S-alkylcysteines in which we were only interested as a parameter for exposure to alkylating agents.

MATERIALS AND METHODS

Chemicals used were of the analytical quality mentioned before[13].

Urine samples were stored and deproteinized if necessary, analysis of free SH-groups, hydrolysis and estimation of creatinine was performed as in our original test[13]. The reduction of disulphides was performed with sodium borohydride according to the method of Cavallini[14], modified according to Modig[15].

Determination of SH-groups

Free SH-groups after reduction with $NaBH_4$: 1 ml of clear urine or of supernatant after deproteinization was mixed with 2 drops of an antifoam solution (n-octanol/ethylalcohol 1 : 9 v/v) and 1 ml of a freshly prepared solution of 5% $NaBH_4$ in distilled water (w/w). The reactionmixture was heated at 60° for 15 min., after which the excess of $NaBH_4$ was destroyed with 0.5 ml 2.7M HCl. The reactionmixture was cooled with occasional shaking during 10 min. Then 0.2 ml metaphosphoric acid was added. The solution was mixed and after 5 min. 0.25 ml of it was mixed with 0.3 ml of the DTNB reagent and 2 ml of the buffer mentioned before. Again a blank for the urine was run with distilled water, instead of the DTNB reagent, a blank for the reagent with distilled water instead of the solution in which the reduction had been performed.

After reading the extinction at 412 nm between 1 and 6 min. the number of mMol SH eq./1 was calculated as:

$$\frac{2.7}{1} \times \frac{2.55}{0.25} \times \frac{E_{det.} - (E_{bl.urine} + E_{bl.reagent})}{\text{mol. ex. coeff.}}$$

SH-groups after reduction of -S-S- and alkaline hydrolysis of -S-:
The remainder of the solution which had been treated with $NaBH_4$ was mixed with 1.0 ml 4M NaOH in a dark screw cap vial, closed under nitrogen and heated at 100° for 50 min., then the solution was cooled in ice (10 min.) and neutralized with 1 ml 4M HCl. After 5 min. 0.25 ml of it was mixed with 0.3 ml DTNB solution and 2 ml of the buffer. A blank for the urine was run with distilled water instead of the colouring reagent, a blank for the colouring reagent with distilled water instead of the solution containing the urine.

After reading the extinction at 412 nm, the number of mMol SH eq./1 was calculated as:

$$\frac{2.7}{1} \times \frac{4.2}{2.2} \times \frac{2.55}{0.25} \times \frac{E_{det.} - (E_{bl.urine} + E_{bl.reagent})}{\text{mol. ex. coeff.}}$$

The difference between the latter and the former calculated amounts is the number of mMol SH eq./1 set free from thioethers.

RESULTS AND DISCUSSION

24h Urines were collected from 5 male volunteers, 2 smokers and 3 non-smokers, during 5 successive days and both the urinary S-alkylsulphide and dialkyl-disulphide content was analysed. The results are represented in table 1. Again smokers showed a strongly significant higher thioether excretion than non-smokers. The disulphide excretion, however, appeared as a relative constant value, which had to be expected as an exogenous relatively uninfluenced background contribution. This constant value is in the same order of magnitude as that given by Bir et al.[16]. According to these authors women should show a somewhat lower excretion (± 25 μMol SH eq./ mMol creatinine).

Table 1

EXCRETION OF μMOL SH EQ./mMOL CREATININE IN VOLUNTEERS ($\bar{x}$ ± S.D.)

	-S-S-	-S-	
Smokers:	38 ± 5	33 ± 8	n = 5
	41 ± 3	28 ± 5	
Non-smokers:	36 ± 2	10 ± 3	n = 5
	40 ± 5	12 ± 9	
	38 ± 1	16 ± 11	

The reduction of the disulphide bond:

It has been demonstrated by Fashold et al.[17] that the disulphide bonds are exclusively attacked by the $NaBH_4$-reduction, which is a reason for its abundent use in peptide-chemistry[18,19,20]. To control its inability to decompose thioethers we performed the reduction with benzyl- and iso-propylmercapturic acids: No free SH-groups could be detected after treatment with $NaBH_4$.

The optimal reduction time of 15 min. at 60° was decuded from experiments with oxidized glutathione and cystine. To remove the excess $NaBH_4$ Cavellini[14] used acetone, which, however, gives a precipitate with a urine sample. This does not occur with the method of Modig[15], who used hydrochloric acid for this purpose.

REFERENCES

1. Documenta Geigy, Wissenschaftliche Tabellen. 6. Auflage. Ed. K. Diem, Basel, Switzerland.
2. Ohmori, S., Shimomura, T., Azumi, T. and Mizuhara, S. (1965) Biochem. Z. 343, 9.
3. Kuwaki, T. and Mizuhara, S. (1966) Biochem. Biophys. Acta 115, 491.
4. Rogers, K.M. and Barnsley, E.A. (1977) Xenobiotica 7, 409.
5. Waring, R.H. (1978) Xenobiotica, 8, 265.
6. Turnbull, L.B., Teng, L., Kinzie, J.M., Pitts, J.E., Pinchbeck F.M. and Bruce, R.B. (1978) Xenobiotica 8, 621.
7. Kodama, H., Ikegami, T. and Araki, T. (1974) Physiol. Chem. & Physics 6, 87.

8. Shih, V.E. and Schulman, J.D. (1969) J. Pediatrics 74, 129.

9. Elce, J.S. and Chandra, J. (1973) Steroids 22, 699.

10. Cagen, L.M., Pisano, J.J., Ketley, J.N., Habig, W.H. and Jakoby, W.B. (1975) Biochim. Biophys. Acta 398, 205.

11. Agrup, G., Falck, B., Kennedey, B.-M., Rorsman, H., Rosengren, A.-M. and Rosengren, E. (1973) Acta Dermatovener 53, 453.

12. Agrup, G., Falck, B., Fyge, K., Rorsman, H., Rosengren, A.-M. and Rosengren, E. (1975) Acta Dermatovener 55, 7.

13. Seutter-Berlage, F., Dorp, H.L. van, Kosse, H.J.J., Hoog Antink, J.M.T. and Wagenaars-Zegers, M.A.P. This volume.

14. Cavallini, D., Graziani, M.T. and Dupré, S. (1966) Nature 212, 294.

15. Modig, H. (1968) Biochem. Pharmacol. 17, 177.

16. Bir, K., Crawhall, J.C. and Mauldin, D. (1970) Clin. Chim. Acta 30, 183.

17. Fasold, H., Gundlach, G. and Turba, F. (1961) Biochem. Z. 334, 255.

18. Brown, W.D. (1960) Biochim. Biophys. Acta 44, 365.

19. Light, A. and Sinha, N.K. (1967) J. Biol. Chem. 242, 1358.

20. Seon, B.B.-K. (1967) J. Biochem. 61, 606.